achieve grade A*

Chemistry

A* Study Guide for INTERNATIONAL GCSE

Frank Benfield

GALORE PARK

www.galorepark.co.uk

Published by Galore Park Publishing Ltd
19/21 Sayers Lane, Tenterden, Kent TN30 6BW
www.galorepark.co.uk

Design and typesetting River Design
Illustration by Ian Moores

Printed by Charlesworth Press, Wakefield

ISBN 978 1 905735 44 0

First published 2012

Details of other Galore Park publications are available at www.galorepark.co.uk

Acknowledgements

The publishers wish to express their thanks to the following agencies for permission to use their images:

The publishers are grateful for permission to use the photographs as follows:
P41 Pascal Goetgheluck/Science Photo Library, P88 Martyn F. Chillmaid/Science Photo Library, P100 Mar Photographics/Alamy, P105 Andrew Lambert Photography/Science Photo Library, P105 Science Photo Library, P106 Andrew Lambert Photography/ Science Photo Library, P106 Paul Silverman/Fundamental Photos/Science Photo Library, P107 Andrew Lambert Photography/ Science Photo Library, P107 Trevor Clifford Photography/Science Photo Library, P129 Andrew Lambert Photography/Science Photo Library, P129 Andrew Lambert Photography/Science Photo Library, P129 Andrew Lambert Photography/Science Photo Library, P129 Andrew Lambert Photography/Science Photo Library, P161 Martyn F. Chillmaid/Science Photo Library

Contents

Introduction

A sound grasp of the facts and principles of chemistry is central to the successful study of science at International GCSE. This is not to imply that chemistry is more important than either physics or biology, but it does bridge these subjects in such a way that a secure grasp of chemistry adds relevance to them and allows them to be better understood. This is particularly relevant to students seeking top grades, at whom this book is targeted.

How is this book organised?

This book follows the structure laid out in the Edexcel International GCSE Chemistry specification (4CH0) and, where necessary and appropriate, includes extra material to fulfil the Cambridge International Examinations International GCSE Chemistry specification (0620).

Material that is only applicable to CIE students is indicated by this symbol.

Each section begins with a list of what you are expected to know for that particular topic. There then follows the material that you need to learn. This has been laid out in a revision-friendly way, using bulleted lists and tables to aid visual learning. Worked examples of calculations are also included. You should work through this material in the way that suits you best. You may choose to read it aloud to yourself or a friend, or to write it out in longhand. The material in this book gives you the absolute essential points of the topic, but remember that to revise effectively you will need to rework it, either mentally or on paper, into concise factual notes that you will be able to remember under exam conditions.

Throughout the book you will find tip boxes that will help you to achieve the A* grade. Some of these apply to all topics; others concentrate on a particularly tricky piece of theory, or something that often causes candidates to trip up.

Towards the end of each section is a review box that acts as a checklist. Once you have worked through the section, check that you can do everything listed in the box. If not, use the page references to refer back to the text and learn that part again.

Each section concludes with a set of practice questions. These are written in exactly the style you will encounter in your exam paper and are design to encourage analysis of ideas which are integral to achieving an A* grade. The marks allocated per question give guidance in measuring achievement. The answers are at the back of the book. Practice makes perfect, so once you have completed the questions in this book, get hold of some actual past papers.

Note: We have chosen to use the following spelling for 'sulfur' with an 'f'. Some exam boards will use the alternative spelling 'sulphur' so therefore you should check prior to your examination to ensure you are using the correct spelling. Key words have been printed in **bold**. You should be sure you understand their meaning.

How will I be assessed?

Edexcel

Edexcel candidates take two exam papers:

- Paper 1 lasts two hours and is worth two-thirds of the overall mark.
- Paper 2 lasts one hour and is worth one-third of the overall mark.

Edexcel candidates are not assessed through coursework.

Successful candidates must meet all three assessment objectives:

AO1   Knowledge and understanding
AO2   Application of knowledge and understanding, analysis and evaluation
AO3   Investigative skills (from June 2013 AO3 will change to: Experimental skills, analysis and evaluation of data methods)

CIE

CIE candidates take **two** papers from the following:

- Paper 1, a multiple choice paper lasting 45 minutes and worth 30% of the overall mark.
  and either
- Paper 2, a core curriculum paper lasting one hour 15 minutes and worth 50% of the overall mark.
  or
  Paper 3, an extended curriculum paper lasting one hour 15 minutes and worth 50% of the overall mark.

Plus **one** of the following, all of which are worth 20% of the overall mark:

- Paper 4, coursework conducted at your school.
- Paper 5, a practical test lasting one hour 15 minutes, conducted at your school.
- Paper 6, a written paper on practical theory lasting one hour.

Successful candidates must meet all three assessment objectives:

A   Knowledge with understanding
B   Handling information and problem solving
C   Experimental skills and investigations

Some help with revision

The most common error is to equate success at revision with the time spent on the task. The following advice should help you to revise effectively:

- Never work for more than 30 minutes at a stretch. Take a break.
- Don't revise all day. Divide the day into thirds (you will have to get up at a decent hour) and work two-thirds of the day at most.
- Always start your revision where you finished your last session. You absorb facts better if you always meet them in more or less the same order.
- Don't revise what you already know. That's like practising a complete piano piece when there is only a short part of it that is causing you problems.
- Annotate your revision notes and make use of coloured highlighting to indicate areas of particular difficulty.
- Do some of your revision with fellow students. The knowledge that others find parts of the syllabus tricky relieves anxiety and other students may well have found an effective way to cope, which they can pass on to you.
- Finally, remember that International GCSE chemistry has no hidden pitfalls. If you learn the facts and understand the principles, you will secure the high grade you deserve.

Section One

1 Principles of chemistry

A States of matter

You will be expected to:

* ★ describe the arrangement, movement and energy of particles in each state of matter
* ★ name the processes that result in changes of state
* ★ describe changes in the arrangement, movement and energy of the particles between states.

The particle model and changes of state

The particle model (see Fig. 1a.01) can be used to describe the behaviour of particles (atoms, ions or molecules) in solids, liquids and gases.

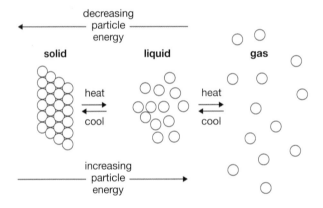

Fig. 1a.01: The particle model

Solid	• Particles are in contact in a fixed arrangement. • Particles do not move from place to place, but vibrate in position.
Liquid	• Particles move randomly while remaining in contact with each other.
Gas	• Particles are not in contact and move randomly. Distances between particles are very much greater than the diameter of the particles. • Even at room temperature the particles in a gas are moving extremely quickly.

Solids, liquids and gases are the three **states of matter**. The changes from one state to another are known as **changes of state**.

The diagram shows the relationship between the three states.

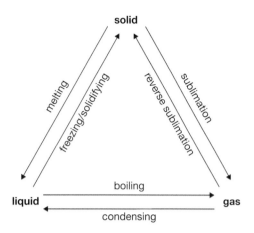

Fig. 1a.02: Changes of state

We can use the particle model to explain changes of state.

- Heating any state of matter makes it expand. The particles move (or vibrate, in the case of solids) more rapidly and so the substance occupies a greater volume, i.e. it **expands**. Cooling causes the reverse effect, **contraction**.
- Heating a solid causes the particles to vibrate more and more violently, until eventually the forces holding them in place are overcome and can no longer hold them in place, so the solid **melts** to form a liquid.
- **Evaporation** takes place at the surface of the liquid when any particle can move fast enough to escape the attractive forces between the particles – it escapes into the air as a gas.

TIP Be careful to distinguish between **boiling** and **evaporation**. Boiling occurs at a fixed temperature (100 °C for water), whereas evaporation occurs at all temperatures, but more rapidly as the temperature rises.

- Heating a liquid causes the particles to move more and more violently until it **boils**, forming gas bubbles in the liquid that escape into the air.
- **Condensation** occurs when the cooling of a gas slows down the movement of the particles, allowing the very small attractive forces between the particles to hold them together sufficiently to form a liquid.
- **Freezing** or **solidifying** occurs when the cooling of a liquid slows the particles down so much that they become held rigidly in place, and so a solid forms.
- A very few substances, such as carbon dioxide, change directly from gas to solid when cooled (this is known as **reverse sublimation**) and solid to gas when heated (this is known as **sublimation**).

TIP A 'model' in scientific terms is a way of simplifying tricky or complex concepts, such as using a diagram to show how atoms and molecules behave in solids, liquids or gases. All models are limited by the assumptions made to simplify them.

The particle model and properties

The particle model helps to explain some of the properties of solids, liquids and gases.

- Solids have a fixed shape and volume, and are difficult to **compress** because the particles are very close together and held in place by strong forces.
- Liquids are only slightly compressible, because there is little free space between the particles.
- Gases can be compressed because the particles are far apart and very weakly held together.
- Gases exert a pressure on the container in which they are placed as a result of the gas particles hitting the container walls – this is easily seen when more gas is added to a balloon and the balloon gets larger.
- Gases and liquids **diffuse** (see 1 B).

Compression in detail

Compression can be explained in terms of the electrons surrounding the atoms of a substance. Compression forces the particles closer together. Electrons all carry the same charge, so the closer they are together the more they repel each other and it gets increasingly difficult to compress the particles.

- The particles in a gas are well separated, so the volume of the gas can be reduced considerably before the repulsive forces become strong enough to prevent further compression.
- In liquids, and particularly in solids, the increasing closeness of the particles means that the repulsion becomes so strong that further compression requires very large forces.

You should now be able to:

★ draw a diagram to show the arrangement of the particles in a solid, liquid and gas (see page 2)
★ describe the movement of the particles in a solid, a liquid and a gas (see page 2)
★ describe the relative energies of the particles in a solid, liquid and gas (see page 2)
★ describe the interconversion of solids, liquids and gases and give the names for the changes involved (see page 3)
★ describe the changes in arrangement, movement and energy of the particles during these interconversions (see page 3).

Practice questions

1. (a) Copy and complete the diagram by naming each of the changes of state taking place. **(4)**

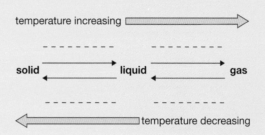

   (b) Describe the process of melting in terms of the particle model. **(3)**

2. Explain the following observations in terms of the particle model.

   (a) You can smell petrol spilt on a garage forecourt a long way from where the spillage occurred. **(3)**

   (b) It is easier to compress air in a syringe than water. **(3)**

   (c) A metal railway rail is longer on a hot summer day than on a cold winter day. **(3)**

3. Copy out the table, writing in the correct word from the two options in each of the nine shaded boxes. **(9)**

Property	Solid	Liquid	Gas
density	decreases/increases ⟶		
compressibility	decreases/increases ⟶		
ease of flow	decreases/increases ⟶		
ability to maintain shape	good/poor	good/poor	good/poor
ability to maintain volume	good/poor	good/poor	good/poor

4. Copy the following paragraphs, completing each blank with the appropriate word.

   When a solid is heated it _____ turning into a _____. Further heating leads to formation of a _____, which returns to a liquid during the process of _____.

   The particles in a gas are _____ apart than those in liquid. In a solid the particles cannot move from place to place, but are able to _____ . **(6)**

B Atoms

Evidence for particles

We can see that particles in liquids and gases are extremely small and in constant, random motion because:

- a coloured solution, when diluted, acquires a uniform colour throughout that becomes increasingly pale the more it is diluted
- small particles such as pollen grains in water or smoke in air are seen to move continuously when viewed under a microscope, as a result of constant random bombardment by the water or air particles respectively.

Diffusion

Particles in liquids and gases are moving in all directions, but there are more particles in an area of high concentration than in an area of low concentration. Diffusion is the **net movement** of particles (the sum of what moves in each direction) from an area of higher concentration to an area of lower concentration. This is sometimes described as moving along (or down) a **concentration gradient**.

- Diffusion can only occur in liquids and gases, where the particles are free to move.
- Diffusion explains why mixtures of liquids and mixtures of gases mix spontaneously and why odours, such as perfume or petrol fumes, can be detected a long way from their source.

Fig. 1b.01: Diffusion

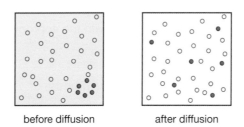

before diffusion   after diffusion

Diffusion at the particle level

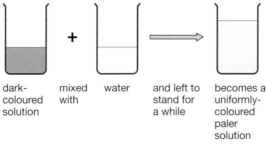

dark-coloured solution   mixed with   water   and left to stand for a while   becomes a uniformly-coloured paler solution

Diffusion in liquids

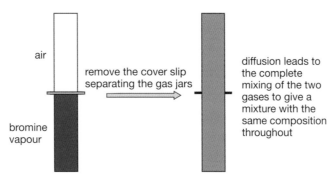

Despite bromine vapour being denser than air, diffusion
leads to complete mixing with the air in the upper gas jar

Diffusion in gases

- Less dense gases diffuse faster than denser ones. This can be demonstrated in the laboratory using the apparatus below. Ammonia (NH_3) is a less dense gas than hydrogen chloride (HCl). The white solid of ammonium chloride (NH_4Cl) forms further away from the ammonia, because the ammonia diffuses faster than the hydrogen chloride.

CAM

> The rate of diffusion is inversely proportional to the square root of molecular mass; gases of low molecular mass diffuse more rapidly than those of higher molecular mass. Ammonia has a molecular mass of 17 g mol^{-1}, hydrogen chloride has a molecular mass of 36.5 g mol^{-1} and the laboratory demonstration shows that ammonia diffuses faster.
>
> $$\frac{Rate_1}{Rate_2} = \sqrt{\frac{M_2}{M_1}} \quad \text{The rate of diffusion of } NH_3 \text{ is } \sqrt{\frac{36.5}{17}} = 1.46 \text{ times faster than HCl}$$

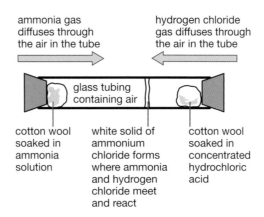

| ammonia gas diffuses through the air in the tube | hydrogen chloride gas diffuses through the air in the tube |

cotton wool soaked in ammonia solution

white solid of ammonium chloride forms where ammonia and hydrogen chloride meet and react

cotton wool soaked in concentrated hydrochloric acid

Fig. 1b.02: Diffusion of ammonia and hydrogen chloride

Atoms, elements, compounds and mixtures

All substances are made of **atoms**. About 100 different types of atom are known.

Elements

- **Elements** are substances that contain just one kind of atom.
- All atoms of an element are chemically identical.
- Atoms of each element are different from one another.
- Atoms of each element have different masses.

Examples of elements

★ hydrogen (H)	★ sulfur (S)	★ sodium (Na)
★ oxygen (O)	★ iron (Fe)	★ chlorine (Cl)

Compounds

- **Compounds** consist of atoms of different elements chemically bonded together – the elements in compounds can only be separated by chemical means.
- All compounds are combinations of atoms of some of the 100 or so known elements.
- Each compound has a constant composition because it contains a fixed ratio of atoms.
- Some compounds are **molecules**, that is atoms joined together by covalent bonds (see Section 1(g)).
- Some compounds do not contain molecules, but are ionic – positive and negative ions attract each other **electrostatically** as a consequence of their opposite charges (see Section 1(f)).
- Compounds usually have properties different from the elements that form them.

Examples of covalent compounds

- ★ methane (CH_4)
- ★ carbon dioxide (CO_2)

Examples of ionic compounds

- ★ sodium chloride (NaCl)
- ★ magnesium oxide (MgO)

> **TIP** The smallest amount of an ionic compound that can be obtained, e.g. NaCl, should not be referred to as a molecule. It is known as an ion pair, Na^+Cl^-.

Mixtures

- **Mixtures** consist of two or more elements or compounds together, but not chemically combined – they can be separated easily.
- A mixture can have any composition, but the composition of a compound is always the same.
- The physical properties of every substance in the mixture are unchanged when they are part of the mixture. This allows differences in physical properties to be used to separate mixtures into the substances of which they are composed.

Examples of mixtures

★ **soil**

★ **sea water**

★ **air**

Alloys

Alloys are mixtures of elements; steel is a mixture of iron and carbon, brass is a mixture of copper and zinc. Alloys differ from other mixtures in two important ways:

- It is difficult to separate alloys into the elements of which they are composed.

- The properties of alloys can be very different from the elements of which they are composed.

For example, steel containing chromium and nickel (18 and 8% respectively) is known as stainless steel, and it does not rust.

Separating mixtures

Substances in a mixture can be separated using the different properties of those substances.

Method	Property used	Used to	Example
filtration	particle size	separate insoluble solids from liquids or solutions	sand from sea water
crystallisation	solubility	separate soluble solids from solutions	salt crystals from sea water
distillation	boiling point	separate pure liquids from solutions	water from sea water
fractional distillation	boiling point	separate two or more liquids that have different boiling points	ethanol from beer
paper chromatography	retention by a third substance, e.g. paper	separate coloured substances	colours in a food dye

Filtration

Filtration works because the large, **insoluble** particles of solid cannot pass through the holes in the filter paper, but the particles of **solvent** and **solution** are small enough to pass through (see Fig. 1b.03).

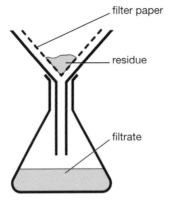

Fig. 1b.03: Filtration

Crystallisation

During **crystallisation** the solvent in a **solution** is evaporated until there is too little left for all the substance to remain dissolved, so some of it appears as solid crystals (see Fig. 1b.04).

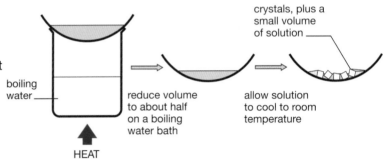

Fig. 1b.04: Crystallisation

Distillation

Water in **aqueous solutions** can be separated from soluble salts by **distillation** (see Fig. 1b.05). The water boils to form water vapour, which is condensed by the cold water in the condenser to liquid water again. The salt has too high a boiling point to evaporate and so is left behind in the flask.

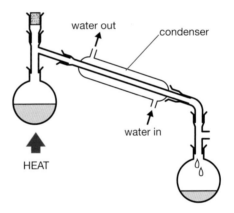

Fig. 1b.05: Distillation

Fractional distillation

Fractional distillation can be used when the two liquids in a mixture have different boiling points (see Fig. 1b.06). For example, water boils at 100 °C but ethanol boils at 78 °C, so, as the temperature is increased the ethanol boils first and distils over into the receiving flask. Only when the temperature is high enough, and all (or most) of the ethanol has distilled over, does the water follow. The receiving flask is removed before the water starts to boil.

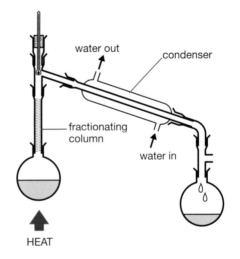

Fig. 1b.06: Fractional distillation

Paper chromatography

In **paper chromatography** solvent is drawn up the paper by **capillary action** carrying the colours placed on the starting line with it (see Fig. 1b.07). The colours are different substances and travel at different speeds. Spots of mixed colour are separated into the individual colours, producing a **chromatogram**. To identify a substance using chromatography you must have the results of a known sample for comparison.

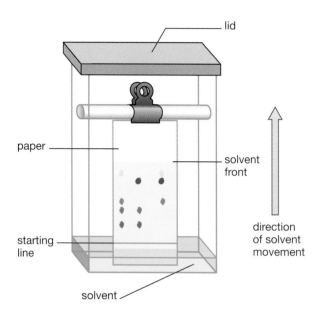

 TIP The starting line on a paper chromatogram should be made in pencil, not ink, which may itself be carried up the paper spoiling the results.

Fig. 1b.07: Paper chromatography

Example

Assume that substance A is the ink from a forged cheque and substances B, C and D are ink samples from the pen of each suspect.

The spots from the pen of suspect C match those in substance A exactly – both samples have the same number of spots in exactly the same places on the paper. Suspect C requires further investigation.

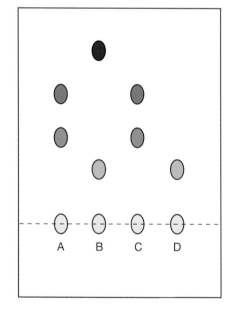

Fig. 1b.08: Chromatogram

CAM

R_f values

Where a particular spot is at the conclusion of a paper chromatography experiment depends on how far the solvent has moved from the starting line, so the same component may well travel a different distance in different experiments. To overcome this problem, and to allow one chromatogram to be compared reliably with another, the R_f value of a component is determined using the formula :

R_f value = distance moved by the component ÷ distance moved by the solvent

The chromatography paper is removed from the container of solvent just before the solvent front reaches the top of the paper, and this point is marked with a pencil line. The paper is then allowed to dry and measured as shown in Fig. 1b.09 to determine the R_f value of each component.

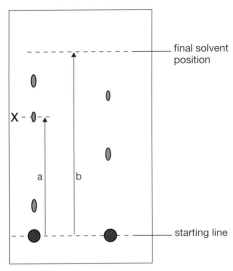

R$_f$ value of spot 'X' = a / b

Fig. 1b.09: Measurements needed to determine R$_f$ values

Paper chromatography is useful in identifying components in mixtures, but it cannot be used to separate a mixture into its pure components. However, more advanced forms of chromatography are used in the pharmaceutical industry to separate the pure drug from impurities formed in the manufacturing process. Removing impurities is important because they may be harmful to the patient, or might reduce the effectiveness of the drug.

Paper chromatography of colourless substances

Mixtures of colourless components can be separated by paper chromatography using **locating agents**. The chromatogram is obtained in the usual way, but after the position of the solvent front has been marked and the chromatogram is dry it is either sprayed with a solution of a locating agent or (if gaseous) placed in a gas jar of it. The individual components react with the locating agent to form coloured compounds.

You should now be able to:

★ describe a simple experiment to show that the particles in solution are small and in constant random motion (see page 6)

★ define the terms *atom* and *molecule* (see page 8)

★ describe differences between elements, compounds and mixtures (see page 8, 9)

★ describe the correct technique to obtain each of the following substances from the mixture containing it: sand from a mixture of sand and water; ethanol from beer; water from sea water; salt from sea water (see page 9, 10)

★ describe how to use paper chromatography to separate a mixture of colours (see page 11).

CAM ★ explain what an R$_f$ value is and how it can be calculated from a chromatogram (see page 11)

★ describe the importance of producing a pure drug from the chemical mixture that is made during drug production (see page 12).

Practice questions

1. Define each of the following terms. **(6)**

   (a) element   (b) compound   (c) mixture

2. Copy the table below and write in the letters in the correct order to get some pure salt crystals from rock salt. **(5)**

first step					last step

   A   Add water   D   Filter the mixture   G   Heat
   B   Crush the rock salt   E   Stir the mixture
   C   Evaporate some of the water   F   Dry the salt crystals

3. Copy and complete the table below. **(15)**

Separation method	Property used	Used to separate	Example
filtration			
crystallisation			
distillation			
fractional distillation			
paper chromatography			

4. Angus wanted to find out whether the red colour of roses and geraniums was due to the same coloured substances. He decided to use chromatography to do this. His teacher told him to grind up some of the petals from each flower with some ethanol to dissolve the coloured substances, and then to filter the mixture before carrying out the chromatography.

   (a) Describe how he should carry out the chromatography experiment **(5)**

   (b) His finished chromatogram is shown in the diagram.

   (i) Why was pencil, rather than a ball-point pen, used to draw the
        starting line? **(2)**

   (ii) Is the red colouring in a rose the same as that in a geranium?
        Explain your answer. **(3)**

   CAM (c) Determine the R_f value for the spot on the chromatogram marked
   with an X. Assume the solvent just reached the top of the paper. **(3)**

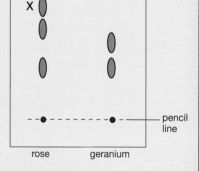

rose   geranium

5. Explain the following in terms of the particle model.

   (a) Two liquids will eventually mix, even if they are not stirred. **(3)**

   (b) The process of diffusion becomes more rapid as the temperature
       increases. **(3)**

C Atomic structure

You will be expected to:

* ★ describe the structure of atoms, including the orbits of electrons
* ★ use the relative masses and charges of protons, neutrons and electrons
* ★ define *atomic number*, *mass number*, *isotope* and *relative atomic mass (A$_r$)*
* ★ calculate the relative atomic mass of an element
* ★ deduce the electronic configurations of elements from their position in the Periodic Table
* ★ deduce the number of outer electrons in an element from its position in the Periodic Table.

The atom

An atom is the basic particle of an element.

* The atom consists of a **nucleus** containing **protons** and **neutrons**. The protons give the nucleus an overall positive charge.
* Negatively charged **electrons** orbit the nucleus. They are held in spherical **shells** by the attraction of the positively charged protons in the nucleus.
* The atom has no overall charge (i.e. it is electrically **neutral**) so, in an atom:

  the number of protons = the number of electrons.

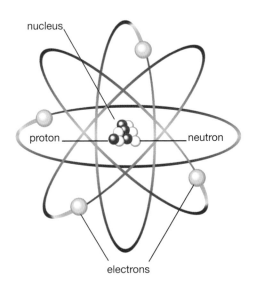

Fig. 1c.01: Atomic structure

Charges and masses of subatomic particles

The mass and charge of each particle are measured relative to those of one neutron.

Particle	Relative charge	Actual mass (kg)	Relative mass
proton (p)	+1	1.67×10^{-27}	1
neutron (n)	0	1.67×10^{-27}	1
electron (e)	−1	9.11×10^{-31}	1/1840

Electron shells

The shells surrounding the nucleus can hold different maximum numbers of electrons.

- Shells are numbered from nearest to the nucleus outwards.
- For the first three shells, electrons generally fill the shells from nearest to the nucleus outwards.

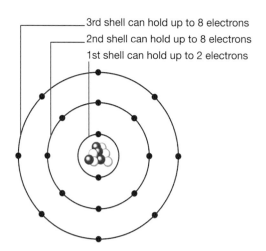

3rd shell can hold up to 8 electrons
2nd shell can hold up to 8 electrons
1st shell can hold up to 2 electrons

Fig. 1c.02: Electron shells

Examples

- Sodium atoms (proton number 11) have 11 electrons: there are 2 in the innermost shell, 8 in the next shell, and 1 in the outermost shell, written as 2.8.1.
- Chlorine atoms (proton number 17) have 17 electrons: there are 2 in the innermost shell, 8 in the next shell, and 7 in the outermost shell, written as 2.8.7.

Working out electronic configurations

Remember that the maximum number of electrons in a shell depends on how close the shell is to the nucleus:

- shell 1 – maximum 2 electrons
- shell 2 – maximum 8 electrons
- shell 3 – maximum 8 electrons up to argon. After Ar the electrons added in K and Ca go into shell 4.

The table below shows the electronic configurations for boron, sodium, sulfur, argon and calcium.

Atomic number	Electrons in				Group	Element
	1st shell	2nd shell	3rd shell	4th shell		
5	2	3			3 (III)	boron
11	2	8	1		1 (I)	sodium
16	2	8	6		6 (VI)	sulfur
18	2	8	8		0	argon
20	2	8	8	2	2 (II)	calcium

TIP Note that for International GCSE the structures of only the first 20 elements (up to calcium) need be known, so only a maximum of eight electrons are found in the 3rd shell, with any remaining in the 4th shell.

Atomic number and mass number

Each element has a unique number of protons in the nucleus.

- The number of protons in an atom is the **atomic number** (also called the **proton number**).
- The number of protons and neutrons in a nucleus is the **mass number** (also called the **nucleon number**).

The atomic number and mass number can be represented by the following shorthand notation:

$$\begin{array}{l}\text{mass number} \quad (A) \\ \text{atomic number} \quad (Z)\end{array} X \quad \text{where X is the element's symbol}$$

> **Example**
>
> hydrogen $^{1}_{1}H$ sodium $^{23}_{11}Na$ carbon $^{12}_{6}C$

Ions

An **ion** is an atom which has had one or more electrons added or taken away. This means it is no longer neutral in charge.

- *Removing* electrons leaves an ion with more protons than electrons so it has a *positive* charge.
- *Adding* electrons leaves an ion with more electrons than protons, so it has a *negative* charge.

> **Examples**
>
> - Sodium ions (Na^+, proton number 11) have 10 electrons: electronic arrangement 2.8.
> - Chloride ions (Cl^-, proton number 17) have 18 electrons: electronic arrangement 2.8.8.

Isotopes

Although the number of protons in a neutrally charged atom equals the number of electrons, neutrons have no charge and can be found in varying numbers in atoms of the *same* element.

Atoms with the same proton number but different numbers of neutrons in their nuclei are called **isotopes**.

- The number of electrons is the same in every isotope of an element.
- The isotopes of an element react chemically in exactly the same way.
- The isotopes of an element differ very slightly in their physical properties.

Example

The element chlorine has the proton number 17. There are two natural isotopes of chlorine.

Isotope	Number of protons	Number of neutrons	Mass number
chlorine-35, ^{35}Cl	17	18	17 + 18 = 35
chlorine-37, ^{37}Cl	17	20	17 + 20 = 37

These isotopes can be represented like this: $^{35}_{17}Cl$ and $^{37}_{17}Cl$

CAM

Radioactive isotopes

Isotopes can be non-radioactive, such as deuterium (hydrogen-2) or oxygen-18, or radioactive, such as uranium-235 or cobalt-60.

The radiation from radioactive isotopes is used to:

- sterilise medical equipment and foodstuffs

- treat cancer

- measure the thickness of paper, cardboard and metal foils

- detect flaws in manufactured components.

The heat given out during nuclear reactions involving uranium-235 and plutonium-239 is used in nuclear power stations to generate electricity.

Calculating relative atomic mass

The **relative atomic mass** (A_r) of an element is measured relative to the mass of a carbon -12 atom, which is given an internationally agreed relative atomic mass of 12.00.

TIP Note that the **relative** atomic mass of an element has no units, but the atomic mass of an element has the unit grams per mole, $g\ mol^{-1}$.

The relative atomic mass of an element is calculated from the relative abundance of its naturally occurring isotopes.

Examples

Naturally occurring chlorine is composed of 75% chlorine-35 and 25% chlorine-37.
Its relative atomic mass equals (0.75 × 35) + (0.25 × 37) = 35.5.

This value appears in the Periodic Table as the relative atomic mass of chlorine.

Magnesium occurs naturally as three isotopes in these proportions: ^{24}Mg 78.6%, ^{25}Mg 10.1% and ^{26}Mg 11.3%.
Its relative atomic mass, to two decimal places, equals (0.786 × 24) + (0.101 × 25) + (0.113 × 26) = 24.33

The Periodic Table

If the elements are listed in order of increasing atomic number an arrangement known as the **Periodic Table** results (see Appendix). Part of it is shown below with the electronic configuration for each element.

TIP

You need to remember, or be able to work out, the electronic configurations of the first 20 elements of the Periodic Table.

Period	Group 1	Group 2	Group 3	Group 4	Group 5	Group 6	Group 7	Group 0
1	H 1							He 2
2	Li 2.1	Be 2.2	B 2.3	C 2.4	N 2.5	O 2.6	F 2.7	Ne 2.8
3	Na 2.8.1	Mg 2.8.2	Al 2.8.3	Si 2.8.4	P 2.8.5	S 2.8.6	Cl 2.8.7	Ar 2.8.8
4	K 2.8.8.1	Ca 2.8.8.2						

These shaded elements lose electrons to reach the more stable condition of a full outer electron shell, as in the inert gases. Those in:
- Group 1 (I) lose 1 electron
- Group 2 (II) lose 2 electrons
- Group 3 (III) lose 3 electrons

These elements are called **metals**.

The elements in Group 4 (IV) neither gain nor lose electrons in reactions, but share them. They are **non-metals**.

These shaded elements gain electrons to reach the more stable condition of a full outer electron shell, as in the inert gases. Those in:
- Group 5 (V) gain 3 (8 – 5) electrons
- Group 6 (VI) gain 2 (8 – 6) electrons
- Group 7 (VII) gain 1 (8 – 7) electrons.

These elements are **non-metals**.

- The elements fall quite naturally into vertical **groups** – atoms of elements in a group have the *same number* of electrons in their *outer shell*.
- Elements are arranged left to right in **periods**. On passing from one element to the next across a period an electron and a proton are added to the atom.
- The number of electrons in the *outer shell* of an atom is equal to the number of the group in the Periodic Table in which the element occurs (except for helium, neon and argon).
- Elements in the same group have very similar properties.
- Group 0 in the Periodic Table consists of elements that show little or no tendency to react – their atoms have *eight* electrons in the outer shells (except helium, which has *two*) which makes them particularly stable. They are known as the **inert** (or **noble**) **gases** (helium, neon, argon, krypton and xenon).
- In reactions, many elements lose or gain sufficient electrons so that their atoms have the same number of electrons as an inert gas.

Chemistry A* Study Guide

You should now be able to:

★ name the particles found in the nucleus of an atom (see page 14)

★ write down the relative charges and masses of the particles present in an atom (see page 14)

★ state where in an atom the electrons are, and how they are held by the nucleus (see page 14)

★ state the meaning of atomic number, mass number, isotope and relative atomic mass (A_r) (see page 16,17)

★ use the fact that bromine has two isotopes, with the mass numbers and abundances: ^{79}Br (50.5 %) and ^{81}Br (49.5%), to calculate the relative atomic mass of bromine (see page 17)

★ write down (a) the electronic configuration and (b) the number of electrons in the outer shell of the elements with the following atomic numbers: 6, 9, 12 and 20 (see page 15)

★ define the terms *group* and *period* as used in the Periodic Table (see page 18)

★ use the fact that an atom of carbon has 6 protons and 6 electrons to calculate the number of protons and electrons in an atom of neon.

Practice questions

1. Define each of the following terms. **(2)**

   (a) atom   (b) ion

2. Copy and complete the table below. **(3)**

Particle	Charge	Relative mass
proton (p)		
neutron (n)		
electron (e)		

3. (a) The diagram below represents an atom of sodium.

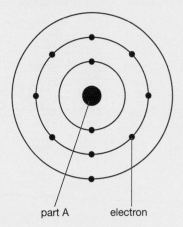

part A   electron

   (i)   Name part A. **(1)**
   (ii)   Which two types of particle does part A contain? **(2)**
   (iii) How are the electrons held in place in the atom? **(2)**

4. If an atom has 10 protons, how many electrons must it have? Justify your answer. **(2)**

5. How many electrons can be held in the 1st, 2nd and 3rd shells in an atom that has 20 or fewer electrons? **(2)**

6. (a) How many electrons are there in the outer shell of:

   (i) a helium atom   (ii) an argon atom? **(2)**

   (b) What is notable about the electronic structures of these two atoms? **(1)**

7. (a) Sodium has a proton number of 11. Write down the electronic structure in shorthand notation
   (i.e. 2.3 etc.) for:

   (i) a sodium atom   (ii) a sodium ion. **(2)**

   (b) In light of your answer to 6(b), explain why the sodium ion has the electronic structure given in 7(a). **(2)**

8. Oxygen atoms have the electronic structure 2.6. Give the symbol for the ion that you would expect oxygen to form. Explain your answer. **(3)**

9. (a) (i) Copy and complete the sentence below by choosing the right word to fill the gaps.

In the symbol $^{40}_{20}Ca$, 40 represents the _____ number of calcium and 20 represents

the _____ number of calcium. **(2)**

(ii) Use the numbers in the symbol to state the number of protons, neutrons and electrons found in calcium. **(3)**

(b) Copy and complete the table below to show the numbers of protons, neutrons and electrons found in the ions shown. **(15)**

ion	protons	neutrons	electrons
$^{1}_{1}H^{+}$			
$^{9}_{4}Be^{2+}$			
$^{56}_{26}Fe^{3+}$			
$^{127}_{53}I^{-}$			
$^{79}_{34}Se^{-}$			

10. Neon has two main isotopes, $^{20}_{10}Ne$ and $^{22}_{10}Ne$.

(a) Explain the term *isotope*. **(2)**

(b) Explain the term *relative atomic mass*. **(2)**

(c) Neon gas is made up of 90% of neon-20 and 10% of neon-22. Calculate the relative atomic mass of neon. **(3)**

(d) How do the chemical properties of each isotope of neon compare with each other? Explain your answer **(3)**

D Relative formula masses and molar volumes of gases

You will be expected to:

★ calculate relative formula masses (M_r)
★ define a *mole*
★ carry out mole calculations using volumes and molar concentrations.

Relative formula mass (M_r)

The **relative formula mass** of a compound is the sum of the relative atomic masses of all the atoms in it. Note that relative formula mass is just a number – it has no units.

> **Worked example**
>
> Calculate the relative formula mass of calcium carbonate, $CaCO_3$. [A_r Ca 40; C 12; O 16].
>
> *Answer*
> The relative formula mass equals:
> 40 (one Ca) + 12 (one C) + 3 × 16 (three O) = 40 + 12 + 48 = 100
>
> The relative formula mass of calcium carbonate is therefore 100.

The mole

The relative atomic mass of any element, expressed in grams, is called one **mole** of the element.

> **Examples**
>
> One mole of these elements has a mass of:
>
> • carbon 12 g   • magnesium 24 g   • fluorine 19 g   • iron 56 g   • lead 207 g
>
> The **mole** can also be defined in terms of number of particles. A mole of anything (atoms, molecules, ions or electrons) contains 6.02×10^{23} of those particles. The number 6.02×10^{23} is called **Avogadro's number**.

 TIP Particular care is needed when dealing with moles of gaseous elements. 'One mole of oxygen' implies one mole of oxygen **molecules** because oxygen occurs naturally as O_2 molecules **not** as atoms. The same applies to H_2, Cl_2, Br_2, F_2 etc. So **one** mole of oxygen molecules, O_2, contains **two** moles of oxygen atoms.
The molar mass of oxygen molecules is $2 \times 16 = 32$ g mol^{-1}

Chemistry A* Study Guide

Using the mole

The mass (in grams) of a compound equal to its relative formula mass is known as the *formula mass of the compound* and contains one **mole** of the compound.

- Formula masses have the unit grams per mole, g mol^{-1}.
- If there is more than one atom of a particular kind in the formula, its relative atomic mass must be multiplied by the appropriate number.

Worked examples

Calculate the mass of one mole of the following:

(i) sodium hydroxide (NaOH):   Na   O   H

                                 23   +   16   +   1   =   **40 g**

(ii) potassium chloride (KCl):   K   Cl

                                 39   +   35.5   =   **74.5 g**

(iii) calcium chloride (CaCl$_2$):   Ca   Cl$_2$

                                 40   +   (2 × 35.5)

                                 40   +   71   =   **111 g**

(iv) iron(II) sulfate (FeSO$_4$):   Fe   S   O$_4$

                                 56   +   32   +   (4 × 16)

                                 56   +   32   +   64   =   **152 g**

(v) ammonium sulfate ((NH$_4$)$_2$SO$_4$):   N   H$_4$ S   O$_4$

                                 2[14 + (4 × 1)]   +   32   +   (4 × 16)

                                 2[18]   +   32   +   64   =   **132 g**

The answers *underlined* represent *one mole* of each of the compounds.

Mole calculations

Starting with the mass of a compound, we can calculate the number of moles it represents, and vice versa.

TIP   Use the triangle in Fig. 1d.01 to help you remember which calculation to use.

Cover up what you want to calculate
eg. Mass = moles x M_r

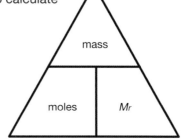

Fig. 1d.01: Molar calculation angle

Worked examples

The relative formula masses for these compounds are taken from the box above.

(i) What fraction of a mole is 8.0 g of sodium hydroxide (NaOH)?

Answer

- one mole of sodium hydroxide has a mass of 40.0 g
- 8.0 g is therefore (8.0 ÷ 40.0) of a mole = **0.20 moles**

(ii) What fraction of a mole is 18.5 g of calcium chloride?

Answer

- one mole of calcium chloride has a mass of 111.0 g
- 18.5 g is therefore (18.5 ÷ 111.0) of a mole = **0.167 moles**

(iii) What is the mass of 0.25 moles of iron(II) sulfate?

Answer

- one mole of iron(II) sulfate has a mass of 152.0 g
- 0.25 moles weighs 0.25 × 152.0 g = **38.0 g**

(iv) What is the mass of 0.80 moles of ammonium sulfate?

Answer

- one mole of ammonium sulfate has a mass of 132 g
- 0.80 moles weighs 0.80 × 132 g = **105.6 g**

Molar volumes of gases

The **molar volume** of a gas is the volume occupied by one mole of gas particles under specified conditions of temperature and pressure.

- At 25 °C and 1 atmosphere pressure, the molar volume = 24 000 cm^3 = 24 dm^3.
- All gases have the *same* molar volume under specified conditions of temperature and pressure.

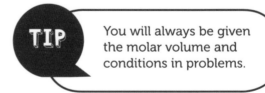

TIP You will always be given the molar volume and conditions in problems.

Calculations using molar volumes of gases

When calculating molar volume, start with the mass, calculate the number of moles and from that calculate the volume.

Worked examples starting with mass

Calculate the volume occupied by (i) 11.0 g of carbon dioxide and (ii) 64.0 g of oxygen. (molar volume at 25 °C = 24 dm³)

Answer

(i) The relative molecular mass of CO_2 = 12 + (2 × 16) = 44

44 g of CO_2 occupy 24 dm³ at 25 °C

hence 11 g occupy (11 ÷ 44) × 24 dm³ = **6.0 dm³**

(ii) The relative molecular mass of O_2 = (2 × 16) = 32

32 g of O_2 occupy 24 dm³ at 25 °C

hence 64 g occupy (64 ÷ 32) × 24 dm³ = **48 dm³**

The reverse of such calculations, i.e. mass of gas from a knowledge of its volume, is straightforward.

Worked examples starting with an equation

Assume throughout that the molar volume at 25 °C = 24 dm³.

(i) Calculate the volume of hydrogen sulfide formed at 25 °C when 22.0 g of iron(II) sulfide react with excess hydrochloric acid.

Answer

FeS   +   2HCl → $FeCl_2$ + H_2S
88   1 mole

22 g of iron(II) sulfide give (22 ÷ 88) × 24 dm³ of H_2S = **6.0 dm³**

(ii) Calculate the volume of hydrogen measured at 25 °C which on burning forms 1.8 g of water.

Answer

$2H_2$ + O_2 → $2H_2O$
2 moles occupying 2 × 24 dm³ at 25 °C   36 g

36 g of water are formed from 48 dm³ of hydrogen

1.8 g of water are formed from (1.8 ÷ 36) × 48 dm³ of hydrogen = **2.4 dm³**

(iii) A piece of zinc was treated with hydrochloric acid until no zinc remained. 250 cm³ of hydrogen measured at 25 °C were collected. Calculate the mass of the zinc.

Answer

Zn   +   2HCl → $ZnCl_2$ + H_2
65   1 mole occupying 24 dm³ at 25 °C

24 dm³ of hydrogen would result when 65 g of zinc react

250 cm³ of hydrogen would result when (65 ÷ 24 000) × 250 g of zinc react = **0.677 g**

You should now be able to:

★ calculate relative formula masses.

★ carry out mole calculations.

Practice questions

1. Iron reacts with dilute sulfuric acid according to the equation:

   $$Fe(s) + H_2SO_4(aq) \rightarrow FeSO_4(aq) + H_2(g)$$

   7.0 g of iron gave 3.0 dm³ of hydrogen at room temperature and pressure when reacted completely with dilute sulfuric acid.

   Calculate the relative atomic mass of iron. [1 mole of any gas occupies 24 dm³ at room temperature and pressure.] **(4)**

2. Iron(III) oxide can be reduced to iron by heating it strongly with carbon:

   $$Fe_2O_3(s) + 3C(s) \rightarrow 2Fe(l) + 3CO(g)$$

   (a) What mass of iron(III) oxide must be reduced to give 7.0 g of iron? **(3)**

   (b) What mass of carbon is required to reduce this mass of iron oxide? **(2)**

   (c) What volume of carbon monoxide is formed during the reduction? **(3)**

   (d) What volume of oxygen is required to convert this carbon monoxide to carbon dioxide?

   $$2CO(g) + O_2(g) \rightarrow 2CO_2(g)$$ **(2)**

3. Lithium nitrate, $LiNO_3$, decomposes on heating according to the equation:

   $$4LiNO_3(s) \rightarrow 2Li_2O(s) + 4NO_2(g) + O_2(g)$$

   [One mole of any gas at room temperature and atmospheric pressure has a volume of 24 dm³. A_r: Li 7; N 14; O 16.]

   (a) A student heated 1.38 g of lithium nitrate and collected the gaseous products in a syringe at room temperature and atmospheric pressure.

      (i)   How many moles of nitrogen dioxide can be obtained from one mole of lithium nitrate? **(1)**

      (ii) What is the relative formula mass of lithium nitrate? **(1)**

      (iii) How many moles of lithium nitrate are there in 1.38 g? **(1)**

      (iv) What mass of nitrogen dioxide can be obtained from the 1.38 g of lithium nitrate? **(1)**

      (v) What volume of nitrogen dioxide can be obtained from the 1.38 g of lithium nitrate? **(1)**

      (vi) What total volume of gas can be obtained from the 1.38 g of lithium nitrate? **(1)**

      (vii) What mass of lithium oxide would remain after 1.38 g of lithium nitrate had been completely decomposed? **(2)**

   (b) Lithium oxide reacts with water according to the equation:

   $$Li_2O(s) + H_2O(l) \rightarrow 2LiOH(aq)$$

   Calculate the concentration of hydroxide ions in solution in mol dm⁻³ when lithium oxide from this experiment (i.e. in (a) (vii)) is dissolved in 250 cm³ of water. **(2)**

E Chemical formulae and chemical equations

You will be expected to:

★ write word equations and balanced chemical equations
★ use state symbols in chemical equations
★ show how the formulae of simple compounds can be obtained experimentally
★ calculate empirical and molecular formulae from experimental data
★ calculate reacting masses using experimental data and chemical equations
★ calculate percentage yield
★ carry out mole calculations using volumes and molar concentrations.

Writing equations

Reactions can be represented using **word equations**, such as:

copper + oxygen $\rightarrow$ copper oxide

copper oxide + hydrogen $\rightarrow$ copper + water

iron + sulfur $\rightarrow$ iron sulfide

- The chemicals that the reaction begins with are the **reactants** (or reagents).
- The chemicals that are formed in the reaction are the **products**.
- Word equations give no information about the masses of the materials involved, nor (obviously) their formulae.

A word equation is changed to a **chemical equation** as follows.

- Replace the name of each substance with its correct formula.
  So: copper + oxygen $\rightarrow$ copper oxide becomes $Cu + O_2 \rightarrow CuO$
- The equation *must* obey the law of conservation of mass.
  In this equation, the reactants are one copper atom and two oxygen atoms.
  The products are one copper atom and one oxygen atom. So the equation must now be balanced.

Balancing symbol equations and the law of conservation of mass

The law of conservation of mass states that in a chemical reaction, the mass of the reactants must equal the mass of the products. This law cannot be broken.

The products of a reaction are made up from exactly the same atoms as the reactants, so their masses must be the same. Also the numbers of atoms of an element on the left and right of the equation must be the same.

To balance a symbol equation you add a **coefficient** (number) *in front* of the chemical formulae so that the law of conservation of mass is obeyed.

- To balance the oxygens on both sides of the equation we can put a 2 in front of the CuO:

  $Cu + O_2 \rightarrow 2CuO$

- But we have now *unbalanced* the copper.

- This can be rectified by putting another 2 in front of the Cu:

  $2Cu + O_2 \rightarrow 2CuO$

- The equation is now balanced.

Changing the **formula** of a substance (by changing the subscript numbers) in a symbol equation in order to balance it is **not** allowed. For example, we cannot write $Cu + O_2 \rightarrow CuO_2$ to balance the equation – the formula of copper oxide is CuO and that does not change.

Worked examples

Write balanced symbol equations for the following word equations.

1. Word equation:

   copper sulfate + sodium hydroxide → copper hydroxide + sodium sulfate

   Answer

   Unbalanced chemical equation

   $$CuSO_4 + NaOH \rightarrow Cu(OH)_2 + Na_2SO_4$$

   Step 1: Balance OH by putting a 2 in front of NaOH:

   $$CuSO_4 + 2NaOH \rightarrow Cu(OH)_2 + Na_2SO_4$$

   The equation is now balanced with 1 Cu, 2Na, 1 SO_4 and 2 OH on each side of it.

2. Word equation:

   aluminium + oxygen → aluminium oxide

   Answer

   Unbalanced chemical equation

   $$Al + O_2 \rightarrow Al_2O_3$$

   Step 1: balance O by putting a 3 in front of O_2 and 2 in front of Al_2O_3:

   $$Al + 3O_2 \rightarrow 2Al_2O_3$$

   Step 2: balance Al by putting a 4 in front of Al:

   $$4Al + 3O_2 \rightarrow 2Al_2O_3$$

   The equation is now balanced with 4 Al and 6 O on each side of it.

Gases and the law of conservation of mass

Consider the following reaction:

calcium carbonate + hydrochloric acid → calcium chloride + carbon dioxide + water

$$CaCO_3(s) + 2HCl(aq) \rightarrow CaCl_2(aq) + CO_2(g) + H_2O(l)$$

A flask of calcium carbonate to which hydrochloric acid has been added *will* have a lower mass after the reaction has finished if the gaseous carbon dioxide escapes, as its mass cannot then be recorded by the balance.

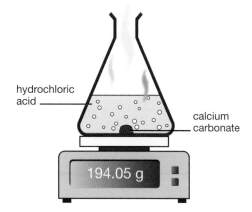

Fig. 1e.01: Calcium carbonate and hydrochloric acid

However, if the carbon dioxide is collected in a syringe or balloon, its mass will be exactly the same as the mass apparently lost during the experiment.

State symbols

State symbols provide additional information about the physical state of the reactants and products under the conditions of temperature and pressure at which the reaction is being carried out.

- (s) means solid
- (aq) means aqueous solution (dissolved in water)
- (g) means gas
- (l) means liquid.

Empirical formulae from experiments

The **empirical formula** shows the simplest whole-number ratio of the atoms present in a substance.

It is possible to calculate empirical formulae either from reacting masses, or from percentage compositions, which have been determined experimentally.

Worked example from reacting masses

0.192 g of magnesium gave 0.320 g of magnesium oxide when burnt in air. Calculate the formula of the oxide formed.

Answer

mass of magnesium oxide = 0.320 g
mass of magnesium = 0.192 g
mass of oxygen in oxide = 0.128 g

1 mole of magnesium has a mass of 24 g: moles of Mg reacted = 0.192 ÷ 24 = 0.008

1 mole of oxygen atoms weighs 16 g: moles of O reacted = 0.128 ÷ 16 = 0.008

The ratio of moles Mg: moles O = 0.008 : 0.008 = 1 : 1, so the formula of the oxide is MgO.

Worked example from percentage composition

A compound contains 28.6% magnesium, 14.3% carbon and 57.1% oxygen. Calculate its empirical formula.

Answer

	magnesium	carbon	oxygen
Divide by A_r	$28.6 \div 24$	$14.3 \div 12$	$57.1 \div 16$
Ratio of atoms	1.19	1.19	3.57
Divide by smallest value (1.19)	1	1	3

Empirical formula is $MgCO_3$

The formula mass is a whole-number multiple of the empirical formula. What the multiple is can only be worked out from a knowledge of the molar mass of the substance.

Molecular formulae

Several compounds may have the same empirical formula: e.g. the empirical formula of both benzene C_6H_6 and ethyne C_2H_2 is CH, but their molecular formulae differ.

- The **molecular formula** shows the *actual number* of each kind of atom present in a molecule of a substance.

- The link between the empirical formula and the molecular formula is the **relative formula mass**.

Worked example

An organic compound has empirical formula CH_2O and a relative formula mass of 120. Calculate its molecular formula.

Answer
Formula mass of the empirical formula: $CH_2O = 12 + (2 \times 1) + 16 = 30$
There must be a whole number of empirical formula units in the molecular formula, call this number n.
$30 \times n = 120$, so $n = 4$
The molecular formula is $(CH_2O)_4 = C_4H_8O_4$

Calculating reacting masses

From a balanced chemical equation, given the mass of one of the chemicals in the reaction, it is possible to work out the mass of any of the other chemicals in the reaction.

Worked example from percentage composition

Mass of product from mass of a reactant

(i) Calculate the mass of magnesium oxide formed when 3 g of magnesium are burned in oxygen.

Answer

$$2Mg \quad + \quad O_2 \quad \rightarrow \quad 2MgO$$

$$(2 \times 24) \qquad\qquad\qquad 2(24 + 16)$$

$$48 \qquad\qquad\qquad\qquad 80$$

48 g of magnesium form 80 g of magnesium oxide
1 g of magnesium forms 80 ÷ 48 g of magnesium oxide
3 g of magnesium form (80 ÷ 48) × 3 g of magnesium oxide = **5.0 g**

(ii) Calculate the mass of hydrogen formed when 0.81 g of sodium react with water.

Answer

$$2Na \quad + \quad 2H_2O \rightarrow 2NaOH \quad + \quad H_2$$

$$(2 \times 23) \qquad\qquad\qquad\qquad (2 \times 1)$$

$$46 \qquad\qquad\qquad\qquad\qquad 2$$

46 g of sodium form 2 g of hydrogen
0.81 g of sodium form (2 ÷ 46) × 0.81 g of hydrogen = **0.035 g**

Mass of reactant from mass of a product

(i) Calculate the mass of calcium carbonate that must be heated in order to make 7 tonnes of calcium oxide.

Answer

$$CaCO_3 \qquad\qquad \rightarrow \quad CaO \quad + \quad CO_2$$

$$40 + 12 + (3 \times 16) \qquad 40 + 16$$

$$100 \qquad\qquad\qquad\qquad 56$$

56 tonnes of calcium oxide are formed from 100 tonnes of calcium carbonate
7 tonnes of calcium oxide are formed from (100 ÷ 56) × 7 tonnes of calcium carbonate = **12.5 tonnes**.

(ii) Calculate the mass of potassium chlorate that must be heated in order to form 3.64 g of potassium chloride.

Answer

$$2KClO_3 \qquad\qquad \rightarrow \qquad 2KCl \qquad + \quad 3O_2$$

$$2[39 + 35.5 + (3 \times 16)] \qquad 2(39 + 35.5)$$

$$245 \qquad\qquad\qquad\qquad\qquad 149$$

149 g of potassium chloride are formed from 245 g of potassium chlorate
3.64 g of potassium chloride are formed from (245 ÷ 149) × 3.64 g of potassium chlorate = **5.99 g**

Mass of a reactant given the mass of another reactant

(i) Calculate the mass of copper(II) oxide that will react with a solution containing 3.60 g of sulfuric acid.

Answer

$$CuO \qquad + \qquad H_2SO_4 \qquad\qquad \rightarrow \quad CuSO_4 \quad + \quad H_2O$$

$$63.5 + 16 \qquad (2 \times 1) + 32 + (4 \times 16)$$

$$79.5 \qquad\qquad\qquad 98$$

98.0 g of sulfuric acid react with 79.5 g of copper(II) oxide
3.60 g of sulfuric acid react with (79.5 ÷ 98.0) × 3.60 g of copper(II) oxide = **2.92 g**

(ii) Magnesium reacts with sulfuric acid to form 3.20 g of magnesium sulfate. What mass of hydrogen was formed at the same time?

Answer

$$Mg \quad + \quad H_2SO_4 \quad \rightarrow \quad MgSO_4 \quad + \quad H_2$$

$$24 + 32 + (4 \times 16) \quad (2 \times 1)$$

$$120 \qquad\qquad\qquad 2$$

120 g of magnesium sulfate are formed together with 2 g of hydrogen
3.20 g of magnesium sulfate are formed together with (2 ÷ 120) × 3.20 g of hydrogen = **0.053 g**

Percentage composition

We can use the relative molecular mass of a compound to calculate the percentage that any element contributes to the composition of the compound. This is useful, for example, when we wish to compare the percentage of nitrogen in different fertilisers such as ammonium nitrate, NH_4NO_3 and urea, $CO(NH_2)_2$.

The mass of one mole of NH_4NO_3 is:
(2×14)[nitrogen] + (4×1)[hydrogen] + (3×16) [oxygen] = 28 + 4 + 48 = 80 g mol^{-1}
so the percentage of nitrogen in ammonium nitrate is $(28 \div 80) \times 100 = 35.0\%$.

A similar calculation for urea gives its molar mass as 60 g mol^{-1}, so the percentage of nitrogen in urea is:
$(28 \div 60) \times 100 = 46.7\%$.

If the price of the two fertilisers were the same, urea would be a better choice as a source of nitrogen.

Worked example

(i) Calculate the percentage of sulfur in potassium sulfate, K_2SO_4.

Answer

First calculate the M_r:

$$K_2 \qquad\qquad S \qquad O_4$$
$$(2 \times 39) \qquad 32 \quad (4 \times 16) \qquad = 174 \text{ g}$$

174 g of potassium sulfate contain 32 g of sulfur
Percentage of sulfur = $(32 \div 174) \times 100\%$ = **18.4%**

(ii) Calculate the percentage of water of crystallisation in hydrated iron(II) sulfate crystals, $FeSO_4.7H_2O$

Answer

$$Fe \qquad\quad S \qquad\quad O_4 \qquad\quad 7H_2O$$
$$56 \quad + \quad 32 \quad + (4 \times 16) + \quad 7(2 + 16)$$
$$56 \quad + \quad 32 \quad + \quad 64 \quad + \quad 126 \quad = 278 \text{ g}$$

278 g of hydrated iron(II) sulfate contain 126 g of water
Percentage of water = $(126 \div 278) \times 100\%$ = **45.3%**

Percentage purity

We sometimes need to know how pure a chemical is, for example when deciding whether a chemical is pure enough to use for a particular purpose, or when valuing scrap metal, where one constituent is worth more than the rest, for example silver in nickel. We do this by carrying out an experiment in which only one substance in the mixture reacts and then comparing the result with that expected had the pure substance been used.

The **percentage purity** is defined as:

$$\frac{\text{actual amount of a chemical present}}{\text{amount of the chemical present if the substance was pure}} \times 100\%$$

Worked example

Magnesium reacts with hydrochloric acid according to the equation below:

$$\text{Mg(s) + 2HCl(aq)} \rightarrow \text{MgCl}_2\text{(aq) + H}_2\text{(g)}$$

A sample of scrap magnesium is contaminated with copper. 1.20 g of the mixture was reacted with excess hydrochloric acid and gave 1080 cm³ of hydrogen (measured at room temperature and pressure) on complete reaction. Calculate the percentage purity of the magnesium.

(A_r Mg 24; molar volume of a gas at room temperature and pressure = 24 dm³ = 24 000 cm³)

Answer

Moles of magnesium used = 1.20 ÷ 24 = 0.05 mol
Moles of hydrogen expected (from the equation) if the magnesium was 100% pure = 0.05 mol
Volume of hydrogen expected if the magnesium was 100% pure = 0.05 × 24 000 cm³ = 1200 cm³
Percentage purity = (volume of hydrogen obtained ÷ theoretical volume of hydrogen expected) × 100%

$$= (1080 \div 1200) \times 100\% = \underline{\textbf{90\%}}$$

Percentage yield

In a reaction, the mass of product you get is called the **yield**.

The **theoretical** yield is calculated from the chemical equation. You may not get as much product as you were expecting from the starting masses of reactants.

The **percentage yield** is defined as:

$$\frac{\text{actual yield (usually in g)}}{\text{calculated theoretical yield (in the same mass units)}} \times 100\%$$

Calculating the percentage yield allows us to choose between different ways of making a particular substance, such as pharmaceutical or fertiliser. As long as there are no major cost differences, the method that gives the highest percentage yield would be chosen.

Worked example

Zinc oxide and sulfuric acid react as follows: $ZnO(s) + H_2SO_4(aq) \rightarrow ZnSO_4(aq) + H_2O(l)$.

In an experiment 4.20 g of zinc oxide were reacted with excess sulfuric acid. The resulting solution of zinc sulfate was evaporated to dryness and the mass of crystals obtained was found to be 6.44 g.

Calculate the percentage yield of zinc sulfate. [A_r Zn 65; S 32; O 16]

Answer

M_r zinc oxide (ZnO) = 65 + 16 = 81   moles ZnO used = 4.20 ÷ 81 = 0.052 mol

From the equation we expect to form the same number of moles of $ZnSO_4$ as the ZnO used = 0.052 mol.

M_r $ZnSO_4$ = 65 + 32 + 64 = 161

Theoretical yield of $ZnSO_4$ = 0.052 × 161 = 8.37 g

Percentage yield = (actual yield ÷ theoretical yield) × 100% = (6.44 ÷ 8.37) × 100% = **76.9 %**

Mole calculations using volumes and molar concentrations

If a known mass of a substance is dissolved in a known volume of water, each cm^3 of the resulting solution contains a known mass of the substance. For example:

- if 10 g of sodium hydroxide are present in 1000 cm^3 (1 dm^3) of solution
- every 1 cm^3 of the resulting solution will contain 10 ÷ 1000 = 0.01 g of sodium hydroxide.

Since the relative formula mass (M_r) of sodium hydroxide = 40:

- 10 g are 10 ÷ 40 = 0.25 mole
- the original 1000 cm^3 of solution therefore contains 0.25 moles of sodium hydroxide
- It is said to be a 0.25 M solution, where the M stands for 'moles per dm^3'.

Worked examples using volumes

1. What mass of sodium hydroxide (M_r 40) would have to be dissolved in 1000 cm^3 of solution to give a 1.0 M solution?

   Answer

   40 g, i.e. 1 mole of sodium hydroxide.

2. What mass of sodium hydroxide (M_r 40) would have to be dissolved in 500 cm^3 of solution to give a 0.25 M solution?

   Answer

   0.25 moles (10 g) of sodium hydroxide dissolved in 1000 cm^3 of solution would give a 0.25 M solution.

   But we only want 500 cm^3 of solution, so half the mass, i.e. 5 g of sodium hydroxide are needed.

A **molar solution** of a substance contains one mole of the substance dissolved in 1000 cm^3 (1 dm^3) of solution.

Solutions of known molarity are prepared by:

- adding a known mass of the substance to a graduated flask (a vessel holding a specified volume of solution when filled to a given mark)
- adding some water and swirling the flask to dissolve the substance completely
- filling with water to the mark
- inverting the flask several times to ensure the contents are properly mixed.

A molar solution of a substance (i.e. a solution of concentration 1.00 mol dm^{-3}) is denoted by the symbol '1 M'; a solution of concentration 0.02 mol dm^{-3} as 0.02 M, etc.

Substance	M_r	Grams in 1 dm^3 of 1 M solution
NaOH	40	40
HCl	36.5	36.5
NaCl	58.5	58.5
H_2SO_4	98	98

Molarities other than 1 M are prepared either by changing the volume of the resulting solution, or by varying the mass of the substance added.

- 40 g of NaOH in 500 cm^3 of solution would give a 2.0 M solution
- 4 g of NaOH in 1000 cm^3 of solution would give a 0.10 M solution
- 4 g of NaOH in 100 cm^3 of solution would give a 1.0 M solution

---

Worked examples using molar solutions

Calculate the molarity of a solution of sodium hydroxide prepared by dissolving 2 g of it in 25 cm^3.

Answer

2 g of sodium hydroxide is 2 ÷ 40 = 0.05 moles
25 cm^3 of solution is 25 ÷ 1000 = 0.025 dm^3

There are 0.05 moles in 0.025 dm^3 so (dividing both sides by 0.025), there are 0.05 ÷ 0.025 = 2.0 moles in 1 dm^3

The resulting solution contains 2.0 moles of sodium hydroxide in 1.0 dm^3 and is therefore a 2.0 M solution.

$$\text{molarity of solution} = \frac{\text{moles of substance dissolved (i.e. mass} \div M_r)}{\text{volume of solution in dm}^3 \text{ OR (volume of solution in cm}^3 \div 1000)}$$

Titration calculations

In titrations we measure the volume of one solution (such as an acid) of a particular molarity which reacts completely with a volume of another solution (an alkali) of a particular molarity. Using the following equation, we can calculate any unknown values:

$$\text{moles of substance} = \frac{\text{volume of solution used (in cm}^3) \times \text{molarity of solution}}{1000}$$

Worked examples

1. Calculate the volume of hydrochloric acid of concentration 0.10 mol dm^{-3} (0.10 M) required to react completely with 25.0 cm^3 of sodium hydroxide of concentration 0.25 mol dm^{-3} (0.25 M)

   Answer

   $NaOH + HCl \rightarrow NaCl + H_2O$

   Moles of NaOH used = molarity × volume in cm^3 / 1000 = 25 × 0.25 × 10^{-3} = 6.25 × 10^{-3}

   From the equation, 1 mole of NaOH reacts with 1 mole of HCl

   So moles of HCl needed to react with 6.25 × 10^{-3} moles NaOH = 6.25 × 10^{-3}

   For the HCl, 6.25 x 10^{-3} = 0.10 × V x 10^{-3} where V is the volume needed.

   So V = **62.5 cm^3**

2. Calculate the concentration (molarity) of a solution of sodium hydroxide, 40.0 cm^3 of which react completely with 10.0 cm^3 of hydrochloric acid of concentration 0.20 mol dm^{-3} (0.20 M).

   Answer

   Moles of HCl used = molarity × volume in cm^3 / 1000 = 0.20 × 10 × 10^{-3} = 2.00 × 10^{-3}

   From the equation, 1 mole of HCl reacts with 1 mole of NaOH

   So moles NaOH needed to react with 2.00 × 10^{-3} moles HCl = 2.00 × 10^{-3}

   For the NaOH, 2.00 × 10^{-3} = M × 40.0 × 10^{-3} where M is the molarity (i.e. the concentration of the NaOH in mol dm^{-3})

   So M = **0.05 mol dm^{-3}**

3. Calculate the volume of sulfuric acid of concentration 0.20 M that will react completely with 40.0 cm^3 of potassium hydroxide of concentration 0.50 M

   Answer

   $2KOH + H_2SO_4 \rightarrow K_2SO_4 + 2H_2O$

   Moles KOH used = 0.50 × 40 × 10^{-3} = 2 × 10^{-2}

   From the equation, 1 mole of KOH reacts with 0.5 mole of H_2SO_4

   So moles H_2SO_4 required = 0.5 × 2 × 10^{-2} = 10 × 10^{-3} = 10^{-2} moles

   For the H_2SO_4, 10^{-2} = 0.20 × V × 10^{-3}

   V = 10 ÷ 0.20 = **50.0 cm^3**

You should now be able to:

★ write and balance chemical equations

★ name the physical state of each substance in the following equation:
$$CaCO_3(s) + 2HCl(aq) \rightarrow CaCl_2(aq) + CO_2(g) + H_2O(l)$$ (see page 29)

★ describe a procedure that allows the formula of magnesium oxide to be determined in the laboratory (see page 29)

★ calculate the empirical and molecular formula of a substance from a knowledge of its percentage composition (see page 30)

★ using the equation that shows how calcium nitrate decomposes on heating:
$$Ca(NO_3)_2(s) \rightarrow CaO(s) + 2NO_2(g) + \tfrac{1}{2}O_2(g)$$
calculate:

    (a) the number of moles of nitrogen dioxide obtained from one mole of calcium nitrate

    (b) the mass of calcium oxide formed from a known mass of calcium nitrate

    (c) the volume of nitrogen dioxide formed from a given number of moles of calcium nitrate

    (d) the total volume of gas formed from a given number of moles of calcium nitrate (see page 26, 31).

Practice questions

1. Use your table of ions to help you to write down the chemical formulae for the following: (10)

   (a) lithium oxide
   (b) calcium bromide
   (c) aluminium oxide
   (d) calcium hydroxide

   (e) ammonium sulfate
   (f) zinc nitrate
   (g) aluminium hydroxide
   (h) silver sulfide

   (i) iron(III) fluoride
   (j) aluminium sulfate

2. Copy and balance the following equations. (20)

   (a) $CaCO_3 + HCl \rightarrow CaCl_2 + CO_2 + H_2O$
   (b) $NaOH + H_2SO_4 \rightarrow Na_2SO_4 + H_2O$
   (c) $K + H_2O \rightarrow KOH + H_2$
   (d) $LiNO_3 \rightarrow Li_2O + NO_2 + O_2$
   (e) $Ca(OH)_2 + HNO_3 \rightarrow Ca(NO_3)_2 + H_2O$
   (f) $CuSO_4 + Al \rightarrow Al(SO_4)_3 + Cu$
   (g) $Al + Fe_2O_3 \rightarrow Al_2O_3 + Fe$
   (h) $NH_3 + O_2 \rightarrow NO + H_2O$
   (i) $NH_4Cl + Ca(OH)_2 \rightarrow NH_3 + CaCl_2 + H_2O$
   (j) $Ca(NO_3)_2 \rightarrow CaO + NO_2 + O_2$

3. Use your table of ions to write down the formulae of the following salts. (10)

   (a) sodium chloride
   (d) potassium sulfide
   (g) potassium carbonate
   (j) iron(III) oxide

   (b) sodium hydroxide
   (e) sodium sulfate
   (h) copper nitrate

   (c) magnesium oxide
   (f) silver nitrate
   (i) iron(II) nitrate

4. Write and then balance chemical equations for the following word equations.

   (a) potassium carbonate + calcium bromide $\rightarrow$ potassium bromide + calcium carbonate
   (b) copper nitrate + sodium hydroxide $\rightarrow$ copper hydroxide + sodium nitrate
   (c) aluminium sulfate + sodium hydroxide $\rightarrow$ aluminium hydroxide + sodium sulfate
   (d) silver nitrate + calcium bromide $\rightarrow$ silver bromide + calcium nitrate
   (e) lead nitrate + aluminium sulfate $\rightarrow$ lead sulfate + aluminium nitrate (15)

5. Calculate the mass of the following. [A_r Cu 64; O 16; S 32; H 1; N 14; Al 27; C 12]

   (a) 0.25 moles of $Al(NO_3)_3$ (2)
   (b) 1.50 moles of $CuSO_4.5H_2O$ (2)

6. 25.00 cm³ of a solution of sodium hydroxide of concentration 0.2 mol dm⁻³ were titrated with sulfuric acid, 20.00 cm³ being required to reach the end-point (i.e. to neutralise the sodium hydroxide exactly).

   (a) How many moles of sodium hydroxide are present in 1 dm³ of solution of concentration 0.2 mol dm⁻³? (1)
   (b) How many moles of sodium hydroxide are present in 25 cm³ of solution of concentration 0.2 mol dm⁻³? (2)
   (c) Use the equation $2NaOH + H_2SO_4 \rightarrow Na_2SO_4 + H_2O$ to help you to calculate how many moles of sulfuric acid are present in the 20 cm³ used in the titration. (2)
   (d) Calculate the concentration of the sulfuric acid in mol dm⁻³. (2)
   (e) Calculate the concentration of the sulfuric acid in g dm⁻³. [A_r H 1, S 32, O 16] (2)

F Ionic compounds

You will be expected to:

★ describe the formation of ions
★ describe oxidation and reduction in terms of the loss or gain of electrons
★ recall and use the charges of common ions
★ work out the charge of an ion from the electronic configuration of its atom
★ use dot and cross diagrams to explain the formation of ions
★ describe ionic bonding in terms of strong electrostatic attraction
★ explain the effect of ionic bonding on the melting and boiling point
★ describe the structure of ionic crystals
★ draw simple diagrams to represent the positions of ions in a crystal.

The formation of ions

The most stable arrangement of electrons occurs when the outer electron shell is complete. This explains why metal atoms *lose* electrons when they form **ions**, and why non-metal atoms *gain* electrons when they form ions. Positively-charged ions are known as **cations**, negatively-charged ions are known as **anions**.

When they form ions:

- Group 1 atoms (lithium, sodium, potassium) lose 1 electron
- Group 2 atoms (beryllium, magnesium, calcium) lose 2 electrons
- Group 3 atoms (boron, aluminium) lose 3 electrons
- Group 5 atoms (nitrogen, phosphorus) gain 3 electrons
- Group 6 atoms (oxygen, sulfur) gain 2 electrons
- Group 7 atoms (fluorine, chlorine) gain 1 electron.

Symbols and charges of some common ions

Cations			Anions		
+1 ions	+2 ions	+3 ions	−1 ions	−2 ions	−3 ions
Li^+	Mg^{2+}	Al^{3+}	F^-	O^{2-}	N^{3-}
Na^+	Ca^{2+}		Cl^-	S^{2-}	
K^+	Zn^{2+}		Br^-		
	Cu^{2+}		I^-		

TIP Remember: Group 4 atoms neither gain nor lose electrons, but share them; and Group 0 atoms don't normally react because their outer electron shell is full.

- Atoms that *lose* electrons become positively charged **cations**:
  e.g. the sodium atom Na [2.8.1] loses 1 electron to form the sodium cation Na⁺ [2.8] (the same electron configuration as Ne)
- Atoms that *gain* electrons become negatively charged **anions**:
  e.g. the chlorine atom Cl [2.8.7] gains 1 electron to form the chloride anion Cl⁻ [2.8.8] (the same electron configuration as Ar)

Oxidation and reduction

- When an atom loses electrons in a reaction, it is **oxidised**: e.g. $Na - e^- \rightarrow Na^+$
- When an atom gains electrons in a reaction, it is **reduced**: e.g. $Cl + e^- \rightarrow Cl^-$
- Oxidation and reduction always occur together in reactions known as **redox reactions**. One reactant loses electrons and is oxidised and the other reactant gains electrons and is reduced.

  For example, in the reaction between magnesium and oxygen $2Mg + O_2 \rightarrow 2MgO$

  two magnesium atoms lose two electrons each: $\qquad 2Mg - 4e^- \rightarrow 2Mg^{2+}$ (**ox**idation)

  two oxygen atoms (in O_2) gain two electrons each: $\qquad O_2 + 4e^- \rightarrow 2O^{2-}$ (**red**uction)

  Adding these equations gives the overall redox reaction: $\qquad 2Mg + O_2 \rightarrow 2MgO$ (**redox**)

Using dot and cross diagrams

Dot and cross diagrams can show how electrons are transferred during the formation of an ionic compound. Dots are used to represent the electrons of one element, and crosses to represent the electrons of the other element.

The dot and cross diagram in Fig. 1f.01 shows the electron transfer during the formation of the salt sodium chloride (NaCl) from sodium and chlorine.

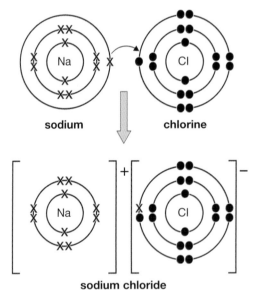

Fig. 1f.01: Dot and cross diagram of formation of NaCl

TIP You may be expected to draw dot and cross diagrams for ionic compounds formed from elements in Groups 1, 2, 3, 5, 6 and 7.

Chemistry A* Study Guide

Ionic bonds

There is a strong **electrostatic attraction** between the oppositely charged ions in an ionic compound. This can produce a large three-dimensional **lattice** structure, which forms **ionic crystals**, in which every cation is surrounded by anions, and *vice versa*.

Fig. 1f.02 shows two ways of representing the three-dimensional arrangement of the ions in NaCl. Sodium chloride crystals are familiar as the crystals of common salt.

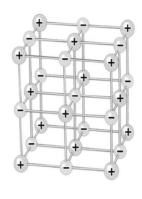

Fig. 1f.02: The three-dimensional arrangement of ions in NaCl

- In sodium chloride each sodium ion is surrounded by six chloride ions and vice versa. This arrangement maximises the total attractive forces between the ions.
- Salt crystals sparkle in the light because the plane (flat) faces of the crystals reflect light well.
- Every sodium chloride crystal has the *same shape*, whatever its size or origin. This external regularity of shape reflects the regular arrangement of the individual ions within the crystal itself.

Fig. 1f.03: Sodium chloride crystals

Properties of ionic compounds

The strong electrostatic attraction between ions in ionic compounds means that a lot of energy is needed to break them, so:

- the crystals are hard
- the compounds have high melting points
- the compounds have high boiling points.

The charges on the ions affects the strength of the attraction between them, and hence their melting and boiling points. For example, magnesium oxide has the same crystal structure as sodium chloride, but because its ions have twice the charge of those in sodium chloride, its melting and boiling points are much higher, as shown in the table below.

Substance	Ions	Melting point (°C)	Boiling point (°C)
NaCl	Na^+ Cl^-	801	1465
MgO	Mg^{2+} O^{2-}	2852	3600

The higher the charges on the ions in an ionic compound, usually the higher the melting point.

- Ionic compounds are always solids at room temperature.
- When melted or dissolved in water the ions they contain are free to move, allowing them to conduct electricity. (The migration of ions to the electrodes during electrolysis is good evidence for the ionic theory of the structure of compounds such as sodium chloride and magnesium oxide.)

You should now be able to:

★ describe how positive and negative ions are formed (see page 39)

★ recall the charges on each of the following ions: e.g. lithium, iodide, oxide, calcium (see page 39)

★ work out the charge on the ion which would be formed from atoms with the following electronic configuration: 2.1; 2.8.7; 2.8.8.2; 2.5 (see page 39)

★ use dot and cross diagrams to explain the formation of calcium ions and chloride ions from atoms of calcium and chlorine (see page 40)

★ understand the nature of the force holding ions together in sodium chloride (see page 41)

★ explain the high melting and boiling points of ionic compounds in terms of their bonding (see page 42)

★ explain why magnesium oxide, MgO, has a higher melting point than sodium chloride, NaCl (see page 42).

Practice questions

1. Use the Periodic Table provided to write down the electronic structures (using the 2.8 notation) of the following: **(10)**

   (a) lithium atom and ion

   (b) oxygen atom and ion

   (c) aluminium atom and ion

   (d) potassium atom and ion

   (e) chlorine atom and ion

2. Copy and complete the diagrams below to show the electronic configuration of an aluminium ion and a chloride ion. **(4)**

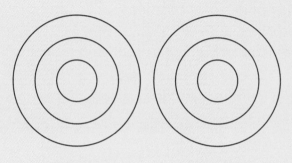

Aluminium ion   Chloride ion

3. Copy and complete the table below to show the charge expected on the ion formed from each atomic electronic configuration. (Note: the letters representing the elements do not represent actual elements.) **(12)**

Symbol for element	Electronic configuration of atom	Electronic configuration of ion	Symbol for ion
A	2.1		
B	2.6		
C	2.7		
D	2.8.2		
E	2.8.3		
F	2.8.8.1		

4. (a) Arrange the following oxides in order of increasing melting point, starting with the lowest: **(2)**

   Al_2O_3 MgO Na_2O.

   (b) Explain the order you have chosen. **(3)**

5. Show the electron transfers which take place to form the ions contained in the following ionic substances: $NaBr$, $MgCl_2$, AlF_3, MgO, Li_3N and K_2O. (Use your table of ions to help you.) **(18)**

G Covalent substances

You will be expected to:

★ describe the formation of covalent bonds

★ explain that a covalent bond involves attraction between shared electrons and the nucleus

★ explain covalent bonding using dot and cross diagrams

★ explain that substances having simple molecular structures may be liquids, gases, or solids with low melting points, at room temperature

★ explain the low melting and boiling points of simple molecular structures

★ explain the high melting points of substances with giant covalent (macromolecular) structures

★ draw simple diagrams of the atoms in diamond and graphite

CAM ★ draw simple diagrams of the atoms in silicon(IV) oxide

★ explain how the uses of diamond and graphite depend on their structures.

Forming covalent molecules

Elements to the *right* of carbon (plus hydrogen) in the Periodic Table can share electrons when they form compounds.

• A shared *pair* of electrons forms a *single* **covalent bond**.

• The result is known as a **molecule** – a number of atoms held together by shared pairs of electrons.

The strong attraction between the negatively charged, shared pairs of electrons and the positive nuclei of the atoms sharing them gives rise to covalent bonds.

The number of electrons that an atom shares depends on how many electrons it has in its outer shell to start with. Each atom shares just enough electrons to give each atom a share in 8 electrons in its outer shell (or, in the case of hydrogen, 2 electrons).

Element	carbon (C)	nitrogen (N)	oxygen (O)	fluorine (F)	neon (Ne)
Number of electrons in outer shell	4	5	6	7	8
Number of electrons needed to give full outer shell	4	3	2	1	0

The electrons from one atom in covalent molecules are often represented by dots and those from the other atoms as crosses in a **dot and cross diagram**.

Examples

In methane, CH_4, carbon 2.4 (i.e. [He].4) requires four more electrons to complete its outer electron shell (like neon). So carbon combines with four hydrogen atoms (each sharing one electron). The carbon has a *share* in 8 electrons (like Ne) while each hydrogen has a share in 2 electrons (like He).

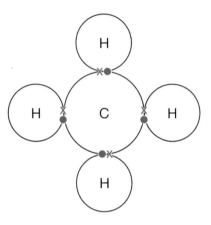

Fig. 1g.01: Dot and cross diagram of methane

In hydrogen chloride, the chlorine shares 1 electron with hydrogen to give it a share in 8, and hydrogen shares 1 electron with chlorine to give it a share in 2.

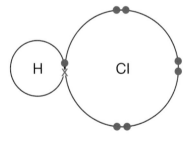

Fig. 1g.02: Dot and cross diagram of hydrogen chloride

CAM

You may be expected to draw a dot and cross diagram for the covalent compound methanol.

TIP

You may be expected to draw dot and cross diagrams for the following covalent compounds: hydrogen, chlorine, hydrogen chloride, water, methane, ammonia, oxygen, nitrogen, carbon dioxide, ethane and ethene.

Simple covalent compounds

Most covalent compounds are simple molecular structures consisting of individual molecules with *strong* covalent bonds between the atoms. However, between the molecules there are *weak* intermolecular forces.

- The weak intermolecular forces give simple covalent compounds *low melting points* – so they could be solids, liquids or gases at room temperature. For example, methane melts at –182 °C and boils at –162 °C.
- Covalent compounds contain no ions and their electrons cannot move because they are firmly held in chemical bonds between the atoms. We can use them as **electrical insulators**, and plastics are very widely used for this purpose.
- Covalent compounds often dissolve in substances like petrol because the weak forces between the molecules can easily be overcome.

TIP Beware: some covalent substances, such as ammonia (boiling point –33 °C), react with water forming ions:

$NH_3(g) + H_2O(l) \rightarrow NH_4^+(aq) + OH^-(aq)$

So their solutions in water **do** conduct electricity.

Giant molecular (macromolecular) substances

In a few covalent compounds, every atom in the structure is joined, via covalent bonds, to every other atom. This produces a giant structure. Examples include diamond, and silicon(IV) dioxide (silicon dioxide).

Diamond

Diamond is formed of carbon atoms covalently bonded with four other carbon atoms into a large, regular, three-dimensional structure. It is very hard and is used in drills and cutting tools.

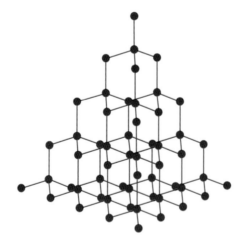

Fig. 1g.03: Structure of diamond

Silicon(IV) oxide (silicon dioxide)

Silicon dioxide is more commonly known as silica or sand. It has a bonding pattern similar to diamond, except that the silicon atoms are not joined directly to each other but to an oxygen atom that forms a 'bridge' between them. This results in a three-dimensional structure similar to diamond. Because all the electrons in diamond and silicon(IV) oxide are used in forming covalent bonds, neither conducts electricity.

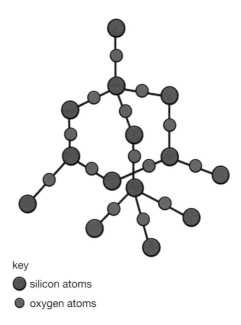

key

● silicon atoms

● oxygen atoms

Fig. 1g.04: Structure of silicon(IV) oxide (silicon dioxide)

Properties of diamond and silicon dioxide

The similar three-dimensional structure of these substances gives them similar properties:

*   very high melting points
*   do not conduct electricity
*   do not dissolve in anything.

Ceramics have structures similar to silicon(IV) oxide and, like it, do not conduct electricity; they are widely used as insulators, particularly for high-voltage overhead power lines.

Graphite

In graphite each carbon atom bonds to three others, using only three of its four electrons. This forms layers of atoms. The remaining electron from each atom is found in the spaces between the layers.

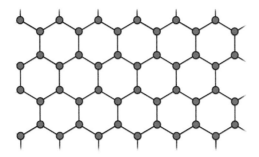

Fig. 1g.05: One layer of graphite seen from above

The **free electrons** can move to carry an electric current. So graphite is a good conductor of electricity. The layers are weakly bonded to, and can slide over, each other. This makes graphite soft, so it is used as a lubricant.

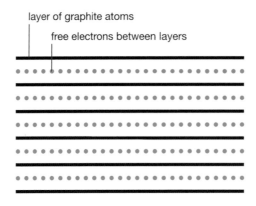

layer of graphite atoms

free electrons between layers

Fig. 1g.06: Layers of graphite seen from the side

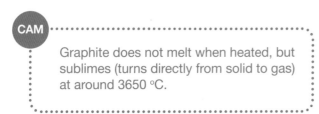

CAM

Graphite does not melt when heated, but sublimes (turns directly from solid to gas) at around 3650 °C.

You should now be able to:

★ describe how covalent bonds are formed between atoms (see page 44)

★ explain how the atoms in a covalent bond are held together (see page 44)

★ draw dot and cross diagrams for the following: H_2, Cl_2, HCl, H_2O, CH_4, NH_3, O_2, N_2, CO_2, C_2H_6 and C_2H_4 (see page 45)

★ explain the low melting points of simple molecular structures in terms of intermolecular forces (see page 46)

★ relate the structures of giant covalent (macromolecular) structures to their properties (see page 46)

★ draw simple diagrams representing the positions of the atoms in diamond and graphite (see page 46, 47)

★ describe the uses made of diamond and graphite and relate them to their structures (see page 46, 48).

Practice questions

1. (a) (i) Explain what we mean when we say that substances have a giant molecular (macromolecular)
        structure.   **(1)**

    (ii) Two giant covalent substances only contain carbon atoms. On the basis of their structures explain
        the differences in the following properties:

    *   electrical conductivity   **(3)**
    *   hardness.   **(3)**

2. Elements form chemical bonds either by losing, gaining or sharing electrons. Illustrate this statement by
    reference to the elements sodium, carbon and chlorine. You should consider the electronic structure of
    each element and its position in the Periodic Table (see Appendix).   **(9)**

3. (a) Draw dot and cross diagrams to show the arrangement of the outer electrons in O_2, HCl, H_2O, NH_3
        and CH_4 **(10)**

    (b) Explain why water, ammonia and methane have the formulae shown, rather than say HO, NH_4 or CH_3 **(4)**

H Metallic crystals

You will be expected to:

★ describe the structure of metals at an atomic level
★ explain some properties of a metal in terms of its structure and bonding.

The structure of a metal

The structure of a metal is best considered as being formed from positive metal ions surrounded by a sea of electrons. The electrons are negatively charged and the ions are positively charged, so they attract each other, holding the structure together.

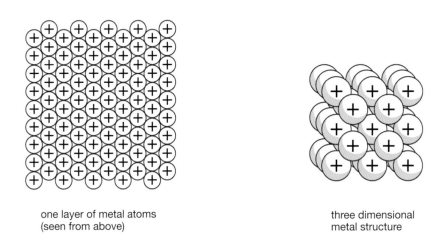

one layer of metal atoms
(seen from above)

three dimensional
metal structure

Fig. 1h.01: The structure of a metal

The properties of metals

The structure of metals has important effects on some of their properties.

- Metals have good **thermal conductivity**.
  The metal ions are in contact so the increase in vibrations caused by heating one part of the metal sample is readily and effectively transmitted to the remainder (see Fig. 1h.02). The sea of electrons also plays a part in this because, as the electrons move through the metal, their energy is also transmitted efficiently to other particles.

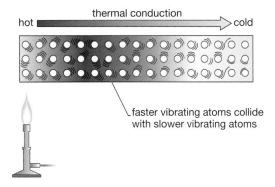

Fig. 1h.02: Thermal conduction in metal

- Metals have good **electrical conductivity**.
  As the electrons in metals are *free to move* and are charged particles, they move easily under the influence of an applied voltage. Their movement in one direction through the metal constitutes an **electric current**.
- Metals show good **malleability**, which means they are easily rolled into sheets.
  Since the metal ions form a series of layers, if sufficient force is applied to one layer it can be made to slide over other layers (see Fig. 1h.03). This changes the thickness and shape of the metal. The metal can therefore be transformed into another shape without losing its mechanical strength.

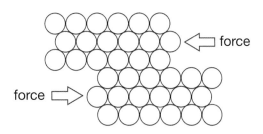

Fig. 1h.03: Changing the shape of metal

You should now be able to:

★ describe the structure of a metal in terms of its ions and electrons (see page 50)
★ explain the malleability and electrical conductivity of a metal in terms of its structure and bonding (see page 51).

Practice questions

1. (a) Sketch the arrangement of the ions and electrons in a metal. (2)

   (b) Explain how the atoms in a metal are bonded to each other. (3)

2. (a) Explain the meaning of the term *malleable*. (2)

   (b) Explain, in terms of the structure of metals, why they are malleable. (3)

   (c) Lithium is the first element in Group 1 of the Periodic Table and francium is the last. Suggest why francium is more malleable than lithium. (3)

3. (a) Explain why metals are good conductors of electricity. (3)

   (b) Suggest why copper (atomic number 29) is a better conductor of electricity than sodium (atomic number 11). (3)

I Electrolysis

Conducting electricity

Only some substances can conduct an **electric current**.

- Ionic substances, if molten or in solution, conduct electricity because the ions are free to move and can carry an electric current.
- Solid ionic substances do not conduct electricity because, although there are ions present, they cannot move through the solid.
- Covalent substances also contain electrons, but they are present in covalent bonds and are unable to move, so no current can flow.

A solution or molten substance that carries the current is known as an **electrolyte**. The passage of an electric current through an electrolyte is called **electrolysis**.

Electrolytes can be identified using a battery connected to an ammeter using the circuit shown in Fig. 1i.01. If the container is filled with an electrolyte, the ammeter will show a current is passing. With a non-electrolyte, no current will pass through the ammeter.

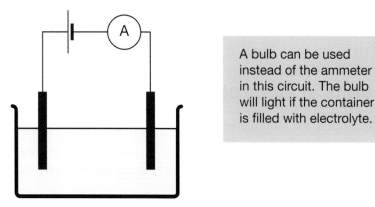

A bulb can be used instead of the ammeter in this circuit. The bulb will light if the container is filled with electrolyte.

Fig. 1i.01: A simple electrolysis circuit diagram

Cations and anions

Electrolysis is easier to understand if the role of the battery is clear in your mind. In a battery, two types of reaction take place:

- one reaction gives up electrons as it occurs
- the other reaction takes up electrons.

The **battery** (or a direct-current power supply) is an electron pump, pushing electrons out at one end and taking them up at the other (see Fig. 1i.02).

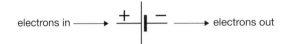

electrons in ⟶ **+** **−** ⟶ electrons out

Fig. 1i.02: The battery as an electron pump

If the battery is connected to a pair of **electrodes** (electrical conductors) immersed in an electrolyte, the ions are attracted towards the electrode with the *opposite* sign (see Fig. 1i.03).

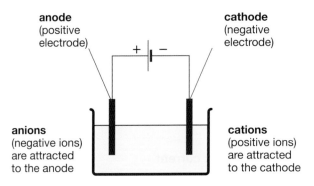

anode
(positive
electrode)

cathode
(negative
electrode)

anions
(negative ions)
are attracted
to the anode

cations
(positive ions)
are attracted
to the cathode

Fig. 1i.03: An electrolytic cell

It is important that the electrodes do not react with the electrolyte or with any of the products formed at the electrodes. Electrodes which meet these requirements are described as **inert electrodes**. Rods of graphite are commonly used as electrodes in the laboratory because, not only are they inert, they also have a high melting point. This means they can be used in the electrolysis of molten salts, such as lead(II) bromide.

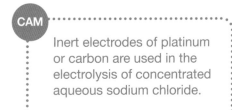

CAM

Inert electrodes of platinum or carbon are used in the electrolysis of concentrated aqueous sodium chloride.

TIP

Remember:

- cations (**positive** ions) move to the cathode (**negative** electrode)
- anions (**negative** ions) move to the anode (**positive** electrode).

Reactions in electrolysis

During electrolysis, the electrolyte is decomposed into simpler substances.

It is important to understand that no chemical reactions take place until the ions actually *reach* the electrodes. In the region between the electrodes the ions only *move* towards the electrodes as a result of their opposite charge.

- At the cathode, cations *gain* electrons, forming atoms, e.g. sodium ion + electron → sodium atom.
- At the anode, anions *lose* electrons, forming atoms, e.g. chloride ion – electron → chlorine atom.

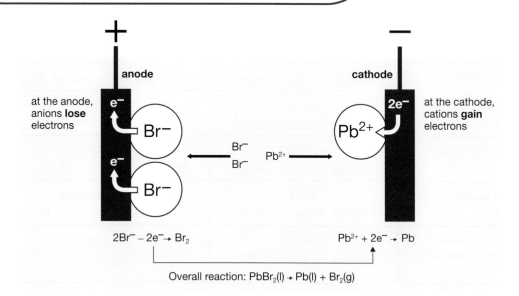

at the anode, anions **lose** electrons

at the cathode, cations **gain** electrons

$2Br^- - 2e^- \rightarrow Br_2$ $Pb^{2+} + 2e^- \rightarrow Pb$

Overall reaction: $PbBr_2(l) \rightarrow Pb(l) + Br_2(g)$

Fig. 1i.04: Electrolysis of PbBr$_2$

- During electrolysis, **oxidation** occurs at the anode and **reduction** occurs at the cathode.
- Metals (and hydrogen) are formed at the cathode, non-metals are formed at the anode.

Using half-equations

Equations that include the electrons are known as **ionic half-equations** because they cannot occur alone. They must take place at the same time as another half-equation in the reaction so that the electrons involved cancel. For example:

- $2Br^- - 2e^- \rightarrow Br_2$
- $Pb^{2+} + 2e^- \rightarrow Pb$

Half-equations must be *balanced* so that they produce the correct product in the overall reaction. So the product at the anode for this reaction must be shown as Br_2 not as Br.

The overall equation is an example of a **redox** reaction, in which both reduction and oxidation occur.

TIP You need to know the products of the electrolysis of these substances.

Electrolyte	Product at cathode and cathode half-equation	Product at anode and anode half-equation
dilute aqueous sodium chloride*	hydrogen $2H^+(aq) + 2e^- \rightarrow H_2(g)$	oxygen $4OH^-(aq) - 4e^- \rightarrow O_2(g) + 2H_2O(l)$
concentrated aqueous sodium chloride*	hydrogen $2H^+(aq) + 2e^- \rightarrow H_2(g)$	chlorine $2Cl^-(aq) - 2e^- \rightarrow Cl_2(g)$
aqueous copper(II) sulfate	copper $Cu^{2+}(aq) + 2e^- \rightarrow Cu(s)$	oxygen $4OH^-(aq) - 4e^- \rightarrow O_2(g) + 2H_2O(l)$
dilute sulfuric acid	hydrogen $2H^+(aq) + 2e^- \rightarrow H_2(g)$	oxygen $4OH^-(aq) - 4e^- \rightarrow O_2(g) + 2H_2O(l)$
molten lead(II) bromide	lead $Pb^{2+}(l) + 2e^- \rightarrow Pb(l)$	bromine $2Br^-(l) - 2e^- \rightarrow Br_2(g)$
CAM concentrated hydrochloric acid	hydrogen $2H^+(aq) + 2e^- \rightarrow H_2(g)$	chlorine $2Cl^-(aq) - 2e^- \rightarrow Cl_2(g)$

* Note that the product obtained at the anode when aqueous sodium chloride is electrolysed depends on the concentration of the solution used.

Note also that hydrogen is obtained, rather than sodium. This is because the reaction to form hydrogen $(2H^+(aq) + 2e^- \rightarrow H_2(g))$ occurs more readily than the one to form sodium $(Na^+ + e^- \rightarrow Na)$.

The oxygen formed at the anode results from the oxidation of hydroxide ions present in the water as a result of its self-ionisation: $H_2O(l) \rightarrow H^+(aq) + OH^-(aq)$.

CAM

Electrolysis of Aqueous Copper(II) Sulfate

The products of the electrolysis depend on the electrodes used, as shown in the table below

Electrode material	Product at cathode and cathode half-equation	Product at anode and anode half-equation
graphite	copper $Cu^{2+}(aq) + 2e^- \rightarrow Cu(s)$	oxygen $4OH^-(aq) - 4e^- \rightarrow O_2(g) + 2H_2O(l)$
copper	copper $Cu^{2+}(aq) + 2e^- \rightarrow Cu(s)$	copper ions $Cu(s) - 2e^- \rightarrow Cu^{2+}(aq)$

Electrolysis of Aqueous Copper(II) Sulfate (*continued*)

Copper metal is required in a high state of purity for use in the electronics industry. It is prepared by reducing copper ores and is about 98% pure. The 98% copper is cast into anodes and placed in an electrolyte of copper(II) sulfate with pure copper cathodes. During the electrolysis the anode passes into solution as copper ions,

$$Cu(s) - 2e^- \rightarrow Cu^{2+}(aq),$$

and these move to the cathode where they are reduced to form pure copper,

$$Cu^{2+}(aq) + 2e^- \rightarrow Cu(s).$$

The anode impurities collect on the bottom of the electrolysis cell as a sludge, which is refined to obtain other metals, mainly nickel and silver.

Electroplating

During the purification of copper the cathode becomes coated with copper metal, and the same kind of reaction occurs with other metals below hydrogen in the reactivity series. Articles made of metal can be coated with a thin layer of another metal in this way – the process is known as **electroplating**. Gold and silver are often plated onto articles (for example pen tops, spectacle frames, watch cases and cutlery) made of less valuable metals to improve their appearance and corrosion resistance and steel car bumpers are chromium plated for the same reasons.

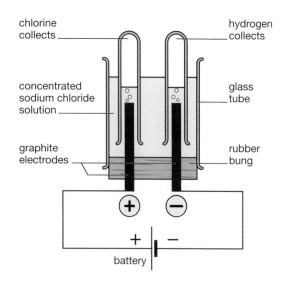

Fig. 1i.05: Apparatus for electrolysis when the products are gases.

Quantitative electrolysis

You need to be able to calculate the mass or volume of a given product of the electrolysis knowing how much of another product was formed. The starting point is to calculate the number of moles involved. The following examples explain how to carry out the calculations.

Worked examples

Example 1

During the electrolysis of molten lead bromide 6.0 dm³ of bromine, measured at room temperature and pressure, was formed at the anode. Calculate the mass of lead formed at the same time.

[Molar volume of a gas at RTP = 24 dm³; A_r Pb = 207]

$$PbBr_2(l) \rightarrow Pb(l) + Br_2(g)$$

6.0 dm³ of bromine = 6.0 ÷ 24 = 0.25 moles

From the equation: 1 mole of lead is formed per mole of bromine.

Moles of lead formed = 0.25 moles

Mass of lead formed = 207 × 0.25 g = **51.8 g**

Example 2

Aluminium is made by the electrolysis of aluminium oxide dissolved in molten cryolite:

$$2Al_2O_3(l) \rightarrow 4Al(l) + 3O_2(g)$$

A factory produces 1000 tonnes of aluminium a week. Calculate the volume of oxygen produced each week.

[1 tonne = 10^6 g; molar volume of a gas at RTP = 24 dm³; A_r Al = 27]

Moles of Al produced each week = 1000 × 10^6 ÷ 27 = 3.70 × 10^7 moles

From the equation: each mole of Al formed is accompanied by the formation of ¾ of a mole of O_2

Moles of O_2 formed = ¾ × 3.70 × 10^7 moles = 2.78 × 10^7 moles

Volume of O_2 formed = 2.78 × 10^7 × 24 dm³ = **6.67 × 10^8 dm³**

Example 3

During the electrolysis of copper sulfate solution the reactions at the anode and cathode are as follows:

anode $4OH^-(aq) - 4e^- \rightarrow 2H_2O(l) + O_2(g)$

cathode $2Cu^{2+}(aq) + 4e^- \rightarrow 2Cu(s)$

During a particular electrolysis 500 cm³ of oxygen was formed at the anode. Calculate the mass of copper deposited on the cathode in the same time.

[Molar volume of a gas at RTP = 24 dm³; A_r Cu = 64]

Moles of O_2 formed = 500 ÷ 24 000 = 0.0208

From the half-equation at the anode, each mole of O_2 requires 4 moles of electrons to form it, but each mole of Cu requires only 2 moles of electrons.

Since the same number of moles of electrons must flow into and out of the electrolysis cell, it follows that moles Cu formed = 2 × moles O_2 formed.

Moles Cu formed = 2 × 0.0208 = 0.0416 moles

Mass Cu deposited on the cathode = 0.0416 × 64 g = **2.66**

Using the faraday in calculations

Some calculations require you to relate the amount of charge passing in an electrolysis circuit to the amount of product formed.

- If a current of 1 amp flows in a circuit for 1 second an electric charge of **1 coulomb** passes around the circuit.
- Approximately 96 500 coulombs of charge equals **1 faraday** of charge.
- 1 faraday of charge corresponds to the transfer of 1 *mole of electrons* around the circuit.

From the above it follows that:

- number of coulombs = current in amps × time in seconds that current flows
- number of faradays = charge in coulombs ÷ 96 500.

One faraday will liberate:

- one mole of a singly-charged ion, e.g. H^+, Na^+, Cl^-
- half a mole of a doubly-charged ion, e.g. Ca^{2+}, Pb^{2+}, O^{2-}
- one-third of a mole of a triply-charged ion, e.g. Al^{3+}.

TIP

Be careful when diatomic gases are being formed.

One faraday liberates **one mole** of atoms from an ion with a single charge, e.g. H^+ or Cl^-. But each mole of the diatomic gas formed contains two atoms (e.g. H_2 or Cl_2). So one faraday liberates half a mole of molecules of gas during electrolysis.

One faraday liberates **half a mole** of atoms from an ion with a double charge, e.g. O^{2-}. But each mole of the diatomic gas formed contains two atoms (e.g. O_2). So one faraday liberates a **quarter of a mole** of molecules of gas during electrolysis.

Worked examples

Example 1

During the electrolysis of molten lead(II) bromide 8.0 g of bromine was formed at the anode. Calculate the mass of lead formed at the cathode during the electrolysis.

$PbBr_2(l) \rightarrow Pb(l) + Br_2(l)$ [A_r Br 80, Pb 207]

Answer

8.0 g of bromine *molecules* = 8 ÷ 160 = 0.05 moles
From the equation, 1 mole of Pb is formed per mole of bromine molecules formed,
so moles Pb formed = 0.05 moles
Mass of Pb formed = moles formed × A_r = 0.05 × 207 g = **<u>10.35 g</u>**

Example 2

One faraday of charge is passed through molten lithium chloride. Calculate (i) the mass of lithium formed and (ii) the volume of chlorine formed. [A_r Li 7, molar volume of a gas at RTP = 24 dm^3]

Answer

(i) The lithium ion is Li$^+$ so 1 faraday will liberate one mole of Li = **7.0 g**

(ii) The chloride ion is Cl$^-$ so 1 faraday will liberate one mole of Cl atoms, which is equal to half a mole of Cl$_2$ molecules.

Volume of Cl$_2$ = moles of Cl$_2$ × 24 dm^3 = **12 dm^3**

Example 3

During the electrolysis of aqueous copper(II) chloride solution 3.175 g of copper was deposited on the cathode. Calculate the volume of chlorine gas formed at the anode at the same time.

$CuCl_2(aq) \rightarrow Cu(s) + Cl_2(g)$ [A_r Cu 63.5, molar volume of a gas at RTP = 24 dm^3]

Answer

Moles of copper formed = 3.175 ÷ 63.5 = 0.05 moles

From the equation, moles of chlorine *molecules* formed = 0.05 moles

Volume of chlorine at RTP = moles of chlorine × 24 dm^3 = 0.05 × 24 = **1.2 dm^3**

You should now be able to:

★ explain (a) why solid sodium chloride and covalent compounds do not conduct electricity, (b) why molten lead bromide and sodium chloride solution do conduct electricity (see page 53)

★ set up and use apparatus in the laboratory to distinguish between an electrolyte and a non-electrolyte (see page 53)

★ describe the electrolysis of aqueous solutions of sodium chloride, copper(II) sulfate and dilute sulfuric acid, and predict the products of electrolysis (see page 56)

★ write ionic half-equations for the reactions taking place at the electrodes during the electrolysis of:
(a) dilute sulfuric acid, (b) aqueous copper(II) chloride (see page 56)

★ calculate the amounts of the products formed during electrolysis from a knowledge of the current passing and the time for which it flows (see page 59).

Practice questions

1. For each of the examples, write the half-equations for the reactions that take place at the electrodes indicated.

   (a) Electrolysis of dilute sulfuric acid: (i) anode, (ii) cathode **(4)**

   (b) Electrolysis of saturated aqueous sodium chloride (brine): anode **(2)**

   (c) Electrolysis of aqueous copper(II) sulfate: cathode **(2)**

   (d) Electrolysis of aluminium oxide dissolved in molten cryolite: (i) anode, (ii) cathode **(4)**

2. Molten lead bromide gives lead and bromine when it is electrolysed.

   $$PbBr_2(l) \rightarrow Pb(l) + Br^2(g)$$

   (a) Explain why the lead bromide must be molten before it can be electrolysed. **(2)**

   (b) Copy and complete the blanks in the sentences below:

   Lead will be formed at the _____ electrode

   Bromine will be formed at the _____ electrode **(2)**

   (c) Explain, in terms of electron transfer, how each of the products is formed. Identify whether the process that is taking place in each case is oxidation or reduction. **(6)**

3. (a) Explain clearly how the current is carried round the circuit during the electrolysis of aqueous sodium chloride **(6)**

   (b) Write equations for the reactions taking place at the anode and cathode during this electrolysis. **(4)**

   (c) At which electrode does (i) oxidation and (ii) reduction occur? Explain your answer. **(4)**

4. (a) Write equations for the reactions which occur at the anode and cathode when the following are electrolysed:

   (i) dilute sulfuric acid **(4)**

   (ii) aqueous copper(II) chloride **(4)**

   (iii) dilute aqueous sodium chloride **(4)**

   (iv) molten lead(II) bromide **(4)**

   (b) Suggest how the product at each electrode might be identified. **(6)**

5. Calculate the number of faradays of charge that must be passed through an electrolysis cell in an aluminium smelter to produce 1 tonne of aluminium.

   $Al^{3+} + 3e^- \rightarrow Al$ [A_r Al 27, 1 tonne = 10^6 g] **(4)**

6. In an aluminium smelter 480 dm³ of oxygen was formed at the anode. Calculate the mass of aluminium formed at the cathode at the same time.

   $2O^{2-} - 4e^- \rightarrow O_2(g)$ $Al^{3+} + 3e^- \rightarrow Al(l)$

   [A_r Al 27, molar volume of a gas at RTP = 24 dm³] **(6)**

Section Two

A   The Periodic Table

You will be expected to:

★ define the terms *group* and *period*

★ identify the positions of metals and non-metals in the Periodic Table

★ classify elements as metals or non-metals on the basis of electrical conductivity and the character of their oxides

★ explain why elements in the same group of the Periodic Table have similar chemical properties

★ describe the noble gases (Group 0) as inert gases and explain their lack of reactivity.

The structure of the Periodic Table

In the modern Periodic Table the elements are arranged in order of increasing atomic (proton) number.

- From left to right across a **period** each electron shell is gradually filled with electrons.
- The elements fall into vertical **groups** in which every atom in the group has the *same number* of electrons in its outer shell. Except for helium, the number of electrons in the outer shell equals the number of the group.
- Elements in the same group have very similar properties – they behave as a family of elements.

Metals and non-metals

Position in the Periodic Table

The Periodic Table below shows the first 20 elements, with their electronic configurations. It also shows which of these elements are metals, non-metals and inert gases.

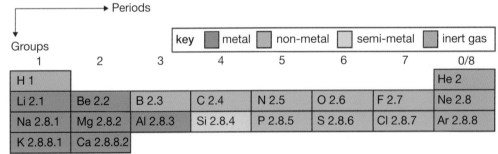

Fig. 2a.01: The first 20 elements of the Periodic Table

TIP   You will not be expected to use the term semi-metal in your answers, but this is the correct classification of silicon because it shows some properties of metals and some of non-metals.

Electrical conductivity in metals and non-metals

The classification of elements as metals or non-metals is based on a study of their properties.

- Metals conduct electricity well. The semi-metal silicon has a small electrical conductivity, but true non-metals (with the exception of graphite) are electrical insulators.
- Metals form basic oxides – when the oxides are added to water the resulting solution has a pH greater than 7. Non-metal oxides form acidic (pH <7) solutions when added to water. The oxide of silicon does not dissolve in or react with water, so on that basis it cannot be classified as either a metal or a non-metal.

Properties of groups

Elements in the same group have the same number of electrons in their **outer** electron shells. As a result they have similar chemical properties. For example:

- Group 1 elements are metals that react with water to give hydrogen and an aqueous solution of their hydroxide
- Group 7 elements react with metals to form ionic solids containing halide ions.

The noble gases (Group 0)

The elements of Group 0, the noble gases, have full outer electron shells. This arrangement is particularly stable, so these gases are very unreactive.

CAM

Argon is present in the air in about one percent by volume and it is the cheapest of the noble gases. Among the industrial uses of argon are:

- to provide an inert atmosphere in filament light bulbs so that the white-hot filament does not oxidise, as it would in air
- to protect reactive metals, such as aluminium from oxidation during welding
- to mix molten metals during alloy formation – argon is blown through the mixture.

The only other noble gas used on a large scale is helium, which has a very low density and is used in airships and meteorological balloons. Neon is used in gas discharge lamps (neon signs) and xenon is used in high-intensity car headlamps and in lasers.

The general trends as you go down a group apply to other groups in the Periodic Table. You may be expected to apply this to examples in other groups in your exam.

You should now be able to:

- ★ define the terms group and period (see pages 64, 18)
- ★ mark the positions of metals and non-metals in the Periodic Table (see page 64)
- ★ name two properties used to classify elements as metals or non-metals (see page 65)
- ★ explain why sodium and potassium have similar chemical properties (see pages 64, 65)
- ★ explain the lack of reactivity of the noble gases (see page 65).

Practice questions

1. The diagrams below show the atoms and molecules present in some elements and compounds.

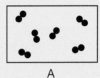
A

B

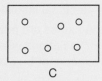
C

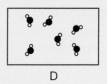
D

   (a) (i) Which diagram could represent the inert gases (Group 0)? Explain your answer. **(2)**

      (ii) Which diagram could represent the halogens (Group 7)? Explain your answer. **(2)**

      (iii) Which diagram could represent a diatomic molecule? Explain your answer. **(2)**

      (iv) Which diagram could represent the product formed when hydrogen and oxygen react? Explain your answer. **(2)**

   (b) The columns of elements in the Periodic Table are known as *groups*. The first five elements in Group 1, and their atomic numbers, are shown in the table below.

Li	3
Na	11
K	19
Rb	37
Cs	55

     (i) How many electrons are there in: a potassium atom, a caesium ion? **(2)**

     (ii) How many protons are there in: a potassium atom, a caesium ion? **(2)**

     (iii) How many electrons are there in the outer shell of each of the elements listed? **(1)**

2. Below is an outline Periodic Table with some elements marked on it. The letters are not the symbols of any particular element.

Period	Group 1	Group 2	Group 3	Group 4	Group 5	Group 6	Group 7	Group 0
1	F							N
2	C		L	M	D	E	J	B
3	H					K	A	
4	I	G						

Answer the following questions by giving the correct letter representing the element. Each letter may be used once, more than once, or not at all.

(a) Which two elements would be the least reactive? **(2)**

(b) Which element would have most electrons in its atom? **(1)**

(c) Which element would have three electrons in its outer shell? **(1)**

(d) Which element would form an ion having a charge of +3? **(1)**

(e) Which element is a flammable gas at room temperature? **(1)**

(f) Which element would be a solid and an electrical insulator? **(1)**

(g) Which element is the most reactive in Group 1? **(1)**

(h) Which element is the most reactive in Group 7? **(1)**

(i) Which element is the most abundant in the atmosphere? **(1)**

(j) Which element would have three more electrons than element C? **(1)**

(k) Identify *two* metals in the same period. **(2)**

(l) Which period contains only non-metals? **(1)**

(m) Identify *two* elements from different groups that form acidic oxides. **(2)**

(n) Identify *two* elements from different groups that form basic oxides. **(2)**

(o) Identify *two* elements from different groups that are gases at room temperature. **(2)**

B Group 1 elements

You will be expected to:

★ describe the reactions of lithium, sodium and potassium with water
★ explain that their reactions with water show them to be a family of elements
★ compare the relative reactivities of the elements in Group 1
★ explain the relative reactivities of the elements in Group 1.

The alkali metals

The elements in Group 1 of the Periodic Table are known as the **alkali metals**.

- They are relatively soft metals with low melting points and low densities.

Metal	Symbol	Melting point (°C)	Boiling point (°C)	Density (g/cm^3)
lithium	Li	180	1330	0.53
sodium	Na	98	890	0.97
potassium	K	64	774	0.86
rubidium	Rb	39	688	1.53
caesium	Cs	29	690	1.90

For comparison, the density of iron is 7.9 g/cm^3.

- Melting point and boiling point decrease as the group is descended. This is because the atoms get bigger down the group and the metalic bond is less strong, so the metal is easier to melt.
- Density increases as the group is descended due to the increasing number of protons in the atom.

Group 1 reactions with water

The Group 1 metals all react with cold water to form hydrogen gas and alkaline solutions of their hydroxides, following the general equation:

$$M(s) + H_2O(l) \rightarrow MOH(aq) + \tfrac{1}{2}H_2(g) \text{ [where M = Li, Na or K]}$$

The similarity of these reactions confirms that lithium, sodium and potassium belong to a family of elements.

Tests for alkalinity

In alkaline solutions: litmus turns blue, phenolphthalein turns pink and Universal Indicator turns purple.

Relative reactivities in Group 1

The reaction with water becomes more vigorous as the group is descended:

- lithium produces a stream of hydrogen bubbles and whizzes around on the surface of the water, but the metal does *not* melt with the heat of the reaction
- sodium melts to form a sphere of metal and whizzes around more rapidly than lithium
- potassium reacts so violently that the metal melts and the hydrogen formed ignites and burns with a lilac flame. (See flame tests in Section 2g.)

This increase in reactivity is due to the increase in size of the metal atom, which makes it easier for the outer electron to be lost.

TIP

Remember: all Group 1 metals have one electron in their outer shell, so they easily **lose** one electron in a reaction.

$$M \text{ (reactive)} \rightarrow M^+ \text{ (stable, inert gas electronic structure)}$$

(where M = Li, Na or K)

This achieves the same electronic structure as the inert gas which precedes the metal in the Periodic Table.

You should now be able to:

★ state what you would see if lithium, sodium and potassium were each added to water, and explain how the observations help to establish these elements as belonging to a family of elements (see page 69)

★ know how the reactivity of the elements in Group 1 changes as the atomic number increases (see page 69)

★ explain why the reactivity of the Group 1 elements changes in the way you have described (see page 69).

Practice questions

1. What is the general trend (increase or decrease) in the following properties of the metals as Group 1 is descended:

   (a) melting point,   (b) density? **(2)**

2. (a) Describe what you would see when a piece of potassium is put into water. **(3)**

   (b) In the reaction between sodium and water name the substance that:

      (i) escapes as a gas **(1)**
      (ii) remains in solution. **(1)**

   (c) Describe how you could show that the substance remaining in solution was alkaline. **(2)**

3. Lithium, sodium and potassium are in Group 1 of the Periodic Table.

   (a) (i) Arrange them in order of increasing reactivity, starting with the least reactive. **(2)**
       (ii) Explain this order of reactivity in terms of their positions in the Periodic Table. **(3)**

   (b) State one *visible difference* between the reactions of lithium and sodium with water. **(1)**

4. A new element has been discovered. Initial studies of the element and its reactions suggest that it should be placed at the bottom of Group 1. Suggest what these studies of the new element might have shown to support this view. **(3)**

C Group 7 elements

The halogens

The elements in Group 7 of the Periodic Table are known as the **halogens**.

They have properties typical of non-metals:

* low melting and boiling points
* do not conduct electricity.

At room temperature, the halogens exist in different states with different colours:

Element	Appearance at room temperature
chlorine	green gas
bromine	red-brown liquid
iodine	grey solid

Group 7 elements all have 7 electrons in their outer shells (one fewer than an inert gas). To increase stability (achieve the structure of the closest inert gas), their atoms share a pair of electrons to form a **diatomic molecule** (contains two atoms). For example, a chlorine molecule is represented as:

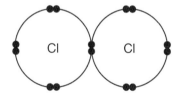

Fig. 2c.01: A chlorine molecule

Reactions of Group 7 elements

Reactions with metals

During reactions with metals, all Group 7 elements gain an eighth electron to form **halide ions** that have the electronic configuration of the *next* inert gas in the Periodic Table.

Halogen atom	
chlorine Cl	
bromine Br	
iodine I	

gains an electron to give

Halide ion	Electronic structure of
chloride ion Cl$^-$	argon
bromide ion Br$^-$	krypton
iodide ion I$^-$	xenon

The reaction with metal forms an ionic salt:

$$2Na + Cl_2 \rightarrow 2NaCl$$

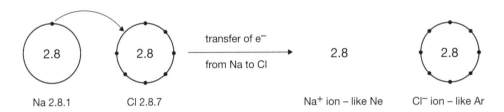

Fig. 2c.02: The reaction of sodium with chlorine

Reactions with hydrogen

All the halogens react with hydrogen to form **hydrogen halides**.

$$H_2 + X_2 \rightarrow 2HX \text{ [where X = Cl, Br or I]}$$

The hydrogen halides are all gases that dissolve in water, reacting with it to give acidic solutions. For example:

$$HCl(g) + H_2O(l) \rightarrow HCl(aq) \rightarrow H^+(aq) + Cl^-(aq)$$

The presence of hydrogen (H$^+$) ions in solution makes the solutions acidic; the hydrochloric acid has **dissociated** to form ions.

An aqueous solution of hydrogen chloride is the acid commonly called *hydrochloric acid*.

hydrogen chloride	HCl
hydrogen bromide	HBr
hydrogen iodide	HI

TIP Be careful to distinguish between:

- hydrogen chloride – a colourless gas
- hydrochloric acid – an aqueous solution of hydrogen chloride.

Chemistry A* Study Guide

Hydrogen chloride in solution

In water

Breaking the H–Cl bond requires energy. In water, this energy is offset by the attraction of the polar water molecules for the H^+ and Cl^- ions formed:

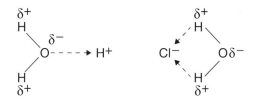

Fig. 2c.03: H⁺ and Cl⁻ ions in water

In methylbenzene

Solutions of hydrogen chloride in methylbenzene are neutral because it is a **non-polar solvent**. The methylbenzene molecules are not attracted to ions in the same way as water, so the energy needed to break the H–Cl bond cannot be offset. The hydrogen chloride remains in solution as *molecules*.

Solutions of hydrogen chloride in methylbenzene do not affect indicators and do not react with metals such as magnesium.

Relative reactivities in Group 7

• The order of reactivity of the halogens is:

  chlorine (most reactive) > bromine > iodine (least reactive)

The elements become *less* reactive both because the positively charged nucleus in larger atoms is shielded from the outer shell by other shells of electrons and because the outer electron shell is further from the nucleus. This means that the nucleus can attract an electron less easily.

Melting point and boiling point increase as the group is descended.

Displacement reactions

A more reactive halogen can displace a less reactive halogen from an aqueous solution of its salt.

For example:

chlorine	+	sodium bromide	$\rightarrow$	bromine	+	sodium chloride
Cl_2	+	$2NaBr$	$\rightarrow$	Br_2	+	$2NaCl$

Bromide ions have been displaced from solution to become bromine.

- Chlorine displaces bromine from solutions of bromides and iodine from solutions of iodides
- Bromine displaces iodine from solutions of iodides.

Colour changes provide evidence that a displacement reaction has taken place:

Note that halide ions in solution are colourless.

Displacement reaction	Observation	Equation
chlorine displacing bromide ions	colourless solution turns orange due to bromine	$Cl_2 + 2Br^- \rightarrow Br_2 + 2Cl^-$
chlorine displacing iodide ions	colourless solution turns red-brown due to iodine	$Cl_2 + 2I^- \rightarrow I_2 + 2Cl^-$
bromine displacing iodide ions	colourless solution turns red-brown due to iodine	$Br_2 + 2I^- \rightarrow I_2 + 2Br^-$

Displacement reactions as redox reactions

The term *displacement* describes what occurs when a more reactive halogen and a less reactive halide ion react, but it gives no clue about what chemistry is taking place.

Displacement reactions can be shown to be redox reactions because

- the more reactive halogen gains electrons to become a halide ion, e.g $Cl_2 + 2e^- \rightarrow 2Cl^-$

- the less reactive halide ion loses electrons to become a halogen, e.g. $2Br^- - 2e^- \rightarrow Br_2$

- adding these two half-equations gives the overall redox reaction, e.g. $Cl_2 + 2Br^- \rightarrow Br_2 + 2Cl^-$

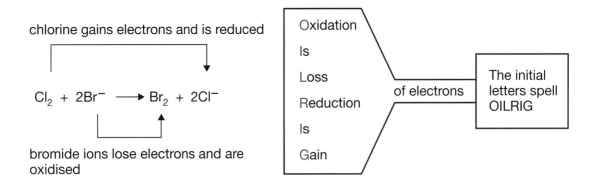

Fig. 2c.04: Halogen redox reaction

You should now be able to:

★ describe the colours and physical states at room temperature of chlorine, bromine and iodine (see page 71)

★ predict the physical state of astatine (the last member of Group 7) at room temperature (see page 71)

★ describe the difference between hydrogen chloride gas and hydrochloric acid (see page 72)

★ explain why aqueous solutions of hydrogen chloride are acidic, while solutions in methylbenzene are not (see page 73)

★ explain how the reactivity of the elements in Group 7 changes with increasing atomic number (see page 73)

★ define the term *displacement reaction* and describe how to carry out a displacement reaction in the laboratory (see page 74).

Practice questions

1. The electronic structures of a sodium atom and a chlorine atom are 2.8.1 and 2.8.7, respectively. When sodium and chlorine react, sodium chloride, NaCl is formed.

   Copy and complete the following statements.

   During the reaction _____ atoms lose electrons to form ions with the symbol _____. **(2)**

   During the reaction _____ atoms gain electrons to form ions with the symbol _____. **(2)**

   The positive ion formed has the same electronic structure as the inert gas _____ and the negative ion formed has the same electronic structure as the inert gas _____. **(2)**

2. (a) The halogens are non-metals. Give two properties typical of non-metals. **(2)**

   (b) Copy and complete the diagram to show the outer electrons in a chlorine molecule. **(3)**

3. Hydrogen burns in chlorine to form a colourless gas.

   (a) Name the colourless gas. **(1)**

   (b) (i) What is a solution of the colourless gas in water called? **(1)**

   (ii) Copy and complete this equation for the reaction of the colourless gas with water to form ions:

   _____ (g) + H_2O(l) → _____ (aq) + _____ (aq) **(3)**

   (c) The colourless gas also dissolves in non-polar solvents, such as methylbenzene.

   Give *three* ways in which you can distinguish between a solution of the colourless gas in water and one in methylbenzene. **(6)**

4. (a) In Group 7, from chlorine to iodine, explain why:

      (i) the atoms increase in size   **(2)**

      (ii) the elements become less reactive.   **(2)**

   (b) Describe an experiment, with the expected result, which would show that chlorine is more reactive than iodine.   **(3)**

   (c) Write a chemical equation, including state symbols, for the reaction that took place in the experiment described in part (b).   **(3)**

5. List the colours and states of the following halogens at room temperature.

   (a) chlorine   (b) bromine   (c) iodine.   **(6)**

6. Copy these sentences, choosing the correct word from each pair.

   The reactivity of the halogens increases/decreases as the atomic number increases.   **(1)**

   The boiling points of the halogens increases/decreases as the atomic number increases.   **(1)**

7. Chlorine will displace bromide ions from a solution of sodium bromide.

   (a) (i) What *type* of reaction is a displacement reaction? Explain your answer.   **(5)**

      (ii) State what you would see when chlorine displaces bromide ions.   **(2)**

   (b) Explain why bromine is less reactive than chlorine.   **(3)**

8. (a) (i) What is meant by the term *polar solvent?*   **(3)**

      (ii) Give one example of such a solvent.   **(1)**

   (b) Explain why gaseous hydrogen chloride forms ions when bubbled into water, but not when bubbled into methylbenzene, a non-polar solvent.   **(4)**

   (c) Describe how you could show that:

      (i) chloride ions are present in the aqueous solution   **(3)**

      (ii) hydrogen ions are present in the aqueous solution.   **(3)**

D Oxygen and oxides

You will be expected to:

* ⋆ state the gases present in air and their approximate percentages by volume
* ⋆ describe experiments to determine the percentage by volume of oxygen in air
* ⋆ describe the laboratory preparation of oxygen from hydrogen peroxide
* ⋆ describe the reactions of magnesium, carbon and sulfur with oxygen
* ⋆ describe the acid-base character of some oxides
* ⋆ describe the laboratory preparation of carbon dioxide from calcium carbonate
* ⋆ describe the formation of carbon dioxide from the thermal decomposition of metal carbonates
* ⋆ understand that carbon dioxide is a dense gas which is moderately soluble in water
* ⋆ explain some uses of carbon dioxide in terms of its properties
* ⋆ describe the formation of acidic solutions when carbon dioxide and sulfur dioxide react with water
* ⋆ describe sulfur dioxide and nitrogen oxides as pollutants that contribute to acid rain
* ⋆ describe problems caused by acid rain.

The gases in air

The atmosphere near the Earth consists approximately of:

* 78% nitrogen
* 20% oxygen
* 1% argon
* 0.03% carbon dioxide and smaller amounts of other substances.

CAM ⋯ These other substances include sulfur dioxide, nitrogen oxides and lead compounds.

Determining the percentage of oxygen in the air

The percentage of oxygen in the air can be estimated using different methods. (see next two pages for examples.)

Oxygen and copper

Oxygen can be removed from air by passing air over heated copper, with which only oxygen reacts, forming solid copper oxide. Nitrogen does not react under these conditions.

The apparatus in Fig. 2d.01 can be used to carry out the experiment.

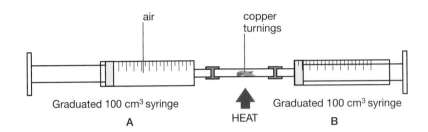

Fig. 2d.01: Apparatus to remove oxygen from the air using copper

- Syringe A is filled with 100 cm³ of air.
- The copper is in the form of turnings to maximise its surface area and enable reaction of all the oxygen.
- The air is passed back and forth between the syringes, over the heated copper turnings, until no further reduction in volume occurs as the copper reacts with the oxygen: $2Cu(s) + O_2(g) \rightarrow 2CuO(s)$.
- The volume of gas remaining will be about 80 cm³, confirming that the air contains about 20% oxygen.

Oxygen and iron

Moist iron wool is placed in the end of a graduated cylinder, which is then inverted and placed in a beaker of water.

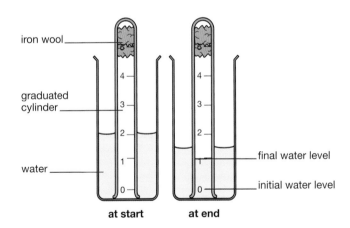

Fig. 2d.02: Apparatus to remove oxygen from the air using iron

- The initial reading on the graduated cylinder is noted.
- The water level in the graduated cylinder rises slowly as oxygen in the air reacts with the iron and the water to form hydrated iron (III) oxide (rust):
  $4Fe(s) + 3O_2(g) + 4H_2O(l) \rightarrow 2Fe_2O_3.4H_2O$
- When the water level ceases to change (after several days) the final reading on the graduated cylinder is recorded.
- The difference between the initial and final readings on the graduated cylinder is the volume of oxygen present in the sample of air. The percentage of oxygen in the air can then be calculated:
  % oxygen = 100 × (initial reading − final reading) ÷ initial reading.

Worked example

In an experiment to measure the percentage of oxygen in the air using iron wool, the following measurements were made:

- initial volume of gas = 220 cm^3
- final volume of gas = 176 cm^3

% oxygen = 100 × (220 − 176) ÷ 220 = 20%

Oxygen and phosphorus

The apparatus for this experiment is a bell jar with a paper scale stuck to it to divide its volume into five equal parts, and a pneumatic trough half-full of water.

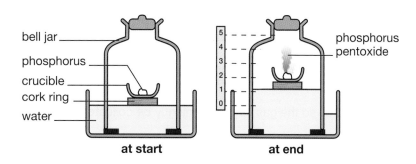

Fig. 2d.03: Apparatus to remove oxygen from the air using phosphorous

- A crucible is placed on a cork ring floating on the water.
- A small piece of white phosphorus is placed in the crucible, which is then quickly covered with the unstoppered bell jar.
- The phosphorus is touched with a hot glass rod to ignite it, and the stopper quickly replaced.
- The burning phosphorus reacts with the oxygen in the bell jar in the reaction:
  $4P(s) + 5O_2(g) \rightarrow 2P_2O_5(s)$.
- The P_2O_5 then reacts with the water to form phosphoric(V) acid:
  $P_2O_5(s) + 3H_2O(l) \rightarrow 2H_3PO_4(aq)$.
- When the phosphorus ceases to burn, it will be seen that about one fifth of the air in the jar has been consumed, showing that air contains about 20% oxygen.

TIP

Of the three experiments, the one involving passing air over heated copper is the most accurate, largely because a syringe can be read with greater precision than the graduations on a measuring cylinder, or the paper scale stuck to the bell jar. Further, the experiments with iron and phosphorus both involved water, in which the gases in air are soluble, and this will lead to inaccurate measurements.

Laboratory preparation of oxygen

Hydrogen peroxide decomposes rapidly in the presence of a **catalyst**, such as manganese(IV) oxide [dioxide], forming oxygen and water:

$$2H_2O_2(aq) \rightarrow 2H_2O(l) + O_2(g)$$

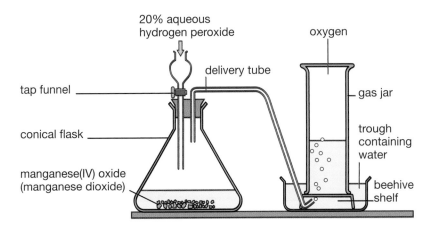

Fig. 2d.04: Apparatus for laboratory preparation of oxygen

Oxygen is only slightly soluble in water, so the gas can conveniently be collected over water.

CAM

Commercial separation of oxygen and nitrogen in the air

The gases present in the air have different boiling points and can be separated in the same way as the fractions in crude oil, by **fractional distillation**.

Air is compressed to 150–200 atmospheres in six stages, cooling at each stage to remove the heat generated by compression. The compressed mixture of gases is then allowed to expand through a fine nozzle, which cools it. These cold gases are then used to cool the compressed air before it reaches the nozzle and the temperature decreases to –200 °C, which is sufficient to liquefy the air.

The liquid air is allowed to warm up in a fractionating column and the nitrogen boils off first at –196 °C, followed by the oxygen at –183 °C thus separating the two. Slightly more complicated fractionation systems can be used to separate the argon and other inert gases.

Reactions of oxygen

Oxygen is a reactive gas and combines chemically with many elements on heating to form **oxides**.

- Magnesium burns fiercely in oxygen with a brilliant white flame to form its oxide, MgO.
  $$2Mg(s) + O_2(g) \rightarrow 2MgO(s) \text{ [white]}$$
- Carbon burns in a plentiful supply of oxygen with a yellow flame to form carbon dioxide.
  $$C(s) + O_2(g) \rightarrow CO_2(g) \text{ [colourless]}$$
- If the oxygen supply is limited, the poisonous carbon monoxide forms.
  $$2C(s) + O_2(g) \rightarrow 2CO(g) \text{ [colourless]}$$
- Sulfur burns in oxygen with a blue flame to form sulfur dioxide, a choking gas.
  $$S(s) + O_2(g) \rightarrow SO_2(g) \text{ [colourless]}$$

The oxides of sulfur and carbon dissolve in water to produce **acidic** solutions.

Magnesium oxide is only slightly soluble in water, but the addition of an indicator to a mixture of the oxide and water shows that the solution is **alkaline**. Magnesium oxide is a basic oxide.

Carbon dioxide

Preparation in the laboratory

Carbon dioxide can be prepared in the laboratory by reacting calcium carbonate with hydrochloric acid:

$$CaCO_3(s) + 2HCl(aq) \rightarrow CaCl_2(aq) + CO_2(g) + H_2O(l)$$

Being denser than air, the gas may be collected by downward delivery (see Fig. 2d.05), or over water (though some will be lost by solution).

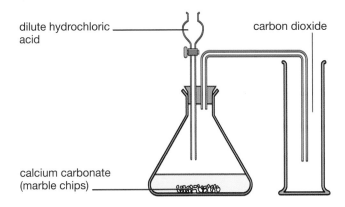

dilute hydrochloric acid

carbon dioxide

calcium carbonate (marble chips)

Fig. 2d.05: Laboratory preparation of carbon dioxide

Thermal decomposition of metal carbonates

Many metal carbonates decompose on heating to form an oxide and carbon dioxide:

$$MCO_3(s) \rightarrow MO(s) + CO_2(g) \quad \text{(where M can represent Ca, Cu or Zn)}$$

The thermal decomposition of calcium carbonate (limestone) is carried out industrially on a huge scale to give calcium oxide (lime),

$$CaCO_3(s) \rightarrow CaO(s) + CO_2(g)$$

Calcium oxide reacts exothermically with water to form calcium hydroxide (slaked lime),

$$CaO(s) + H_2O(l) \rightarrow Ca(OH)_2(s)$$

Calcium hydroxide is sparingly soluble in water and its aqueous solution, common in the laboratory, is known as lime water. Calcium oxide and hydroxide are used to neutralise acidity in soils and to neutralise acidic industrial waste products, such as sulfur dioxide in flue gases (flue gas desulfurisation).

Calcium carbonate is a raw material in the manufacture of cement, made by heating it with clay at 1450 °C in a rotary kiln.

Carbon dioxide and respiration

Respiration is the process in which sugars react with oxygen to produce carbon dioxide and energy. The presence of carbon dioxide in exhaled breath can be easily shown by blowing through lime water, which turns milky.

Properties of carbon dioxide

Carbon dioxide gas is:

- soluble in water – at room temperature 1 dm³ of water dissolves 800 cm³ of the gas
- 1.6 times denser than air.

Reaction with water

Carbon dioxide dissolves in water to form carbonic acid, a very weak acid:

$$CO_2(g) + H_2O(l) \rightarrow H_2CO_3(aq)$$

Uses of carbon dioxide

- A major use of carbon dioxide is in the soft drinks industry. Dissolved carbon dioxide creates the 'fizz' in fizzy drinks. The gas becomes more soluble at greater pressures and at lower temperatures, so the carbon dioxide is added to the cooled drink under pressure. When the bottle or can is opened the pressure is released and bubbles of carbon dioxide come out of the solution. If a cold fizzy drink warms up it goes 'flat' as the carbon dioxide becomes less soluble.
- Carbon dioxide is also used in fire extinguishers. It does not burn and its high density makes it effective at 'blanketing' a fire (excluding oxygen from the area).

Sulfur dioxide

Sulfur dioxide reacts with water to produce an aqueous solution of sulfurous acid:

$$SO_2(g) + H_2O(l) \rightarrow H_2SO_3(aq)$$

Sulfurous acid is a weak acid, but buildings (particularly those built from carbonate rocks such as limestone and marble) can be damaged by constant exposure to sulfurous acid in acid rain.

> **CAM**
>
> Sulfur dioxide is used as a bleach in the manufacture of wood pulp in paper making and to preserve foodstuffs (particularly fruits) by killing bacteria and moulds.

Pollutant gases

Many industrial processes produce pollutant gases. A **pollutant gas** is one that is not normally present in the atmosphere, but which is added as a result of human activity and causes harm to the environment. Pollutant gases include sulfur dioxide and nitrogen oxides.

> **CAM**
>
> Pollutant gases also include carbon monoxide.
>
> In 2000 the EU banned the addition of lead tetraethyl to petrol, so lead compounds is now much less of a pollution problem in the UK. They were phased out and eventually banned because there was evidence that they caused health problems, particularly mental problems in young children.

Producing sulfur dioxide

When fossil fuels such as coal and oil are burnt, the sulfur in them is oxidised to form sulfur dioxide:

$$S(s) + O_2(g) \rightarrow SO_2(g)$$

> **CAM**
>
> Sulfur dioxide can affect health if it is breathed in. In the lungs and other parts of the respiratory system it dissolves in the moisture on surfaces to form acid, which damages tissue. This can cause breathing problems.

Producing nitrogen oxides

We use the term **nitrogen oxides** to cover a mixture of oxides of nitrogen, including nitrogen monoxide and nitrogen dioxide.

When air is subjected to high pressures and temperatures, as happens in internal combustion engines, the nitrogen and oxygen react together.

Nitrogen monoxide is formed first:

$$N_2(g) + O_2(g) \rightarrow 2NO(g)$$

This then reacts with more oxygen to form nitrogen dioxide:

$$2NO(g) + O_2(g) \rightarrow 2NO_2(g)$$

Producing carbon monoxide

Carbon monoxide is formed from the incomplete combustion of fossil fuels. The compete combustion of methane produces carbon dioxide and water:

$CH_4(g) + 2O_2(g) \rightarrow CO_2(g) + 2H_2O(l)$

but when insufficient oxygen is present carbon monoxide is formed:

$CH_4(g) + 1\frac{1}{2}O_2(g) \rightarrow CO(g) + 2H_2O(l)$

The incomplete combustion of liquid hydrocarbons such as hexane in internal combustion engines is the main source of this pollutant.

Acid rain

Sulfur dioxide and the nitrogen oxides are highly soluble in water. For example, sulfur dioxide dissolves in water to form sulfurous acid:

$SO_2(g) + H_2O(l) \rightarrow H_2SO_3(aq)$

These pollutant gases dissolve in water droplets in clouds and fall as **acid rain**.

- Acid rain has caused considerable damage to the environment:
- living organisms, including trees and animals, may be harmed or even killed, and buildings may be damaged by acid **corrosion**.

Gases from power stations (the main source of sulfur dioxide) are now passed through a slurry of calcium hydroxide to neutralise the acidic sulfur dioxide:

$Ca(OH)_2$	+	SO_2	$\rightarrow$	$CaSO_3$	+	H_2O
calcium hydroxide	+	sulfur dioxide	$\rightarrow$	calcium sulphite	+	water

Nitrogen oxides from vehicle exhausts can be removed by catalytic converters that reduce the oxides to nitrogen, for example:

$2NO_2(g) \rightarrow N_2(g) + 2O_2(g)$

The oxides described above follow a general pattern:

- oxides of metals react with water to form alkaline solutions
- oxides of non-metals react with water to form acidic solutions.

There are two other classes of oxide:

Amphoteric oxides. The oxides of some metals, such as aluminium oxide and zinc oxide can show acidic and alkaline behaviour – they are said to be **amphoteric**. These oxides react with both acids and alkalis, dissolving in the process:

Aluminium oxide with dilute hydrochloric acid:
$Al_2O_3(s) + 6HCl(aq) \rightarrow 2Al^{3+}(aq) + 6Cl^-(aq) + 3H_2O(l)$

Aluminium oxide with aqueous sodium hydroxide:
$Al_2O_3(s) + 2NaOH(aq) \rightarrow 2Na^+(aq) + 2AlO_2^-(aq) + H_2O(l)$

The AlO_2^- ion is known as the aluminate ion.

Neutral oxides. These are uncommon, but carbon monoxide (CO), dinitrogen monoxide (N_2O) and water are examples of such oxides. They dissolve in water to form solutions of pH 7.

You should now be able to:

★ name the gases present in air and state their approximate percentages by volume (see page 77)

★ describe an experiment to determine the percentage by volume of oxygen in air (see pages 77, 78, 79)

★ describe the laboratory preparation of oxygen (see page 80)

★ describe what you would see when oxygen reacts with magnesium, carbon and sulfur (see page 81)

★ state the acid-base character of the magnesium oxide, carbon dioxide and sulfur dioxide (see page 81)

★ describe the laboratory preparation of carbon dioxide (see page 81)

★ name the products formed when copper(II) carbonate is heated (see page 82)

★ recall the main physical properties of carbon dioxide (see page 82)

★ explain two uses of carbon dioxide in terms of its properties (see page 82)

★ describe the reactions of carbon dioxide and sulfur dioxide with water to form acidic solutions (see page 82)

★ name two gaseous atmospheric pollutants in acid rain, and describe some problems caused by this pollution (see page 84).

Practice questions

1. Copy, complete and balance the following equations for the reactions of some elements with oxygen.

   (a) $Mg + O_2 \rightarrow MgO$

   (b) $C + O_2 \rightarrow CO$ **(4)**

2. (a) Describe an experiment you could use to determine the percentage of oxygen in air. Include a diagram of the apparatus you would use. **(6)**

   (b) In such an experiment 85 cm³ of air at the start of the experiment had been reduced to 68 cm³ of gas at the end. Calculate the percentage of oxygen in the sample of air. **(2)**

3. (a) What is *acid rain* and how is it formed? **(3)**

   (b) List two harmful effects of acid rain. **(2)**

4. Oxygen can be prepared in the laboratory from hydrogen peroxide.

   (a) (i) Name the catalyst used in the reaction. **(1)**

   (ii) Write a balanced symbol equation for the reaction. **(3)**

5. (a) A piece of burning sulfur is lowered into a gas jar of oxygen.

   (i) Describe what you would expect to see. **(2)**

   (ii) Write a balanced symbol equation for the reaction taking place and name the gaseous product formed. **(3)**

   (b) The gaseous product dissolves readily in water.

   (i) Name the solution formed. **(1)**

   (ii) Describe how you could show that the solution formed was acidic. **(2)**

   (iii) Does your observation in (ii) above indicate that sulfur is a metal or a non-metal? Explain your answer. **(2)**

E Hydrogen and water

You will be expected to:

★ describe the reactions of dilute hydrochloric acid and dilute sulfuric acid with some metals
★ describe the combustion of hydrogen
★ describe the use of anhydrous copper(II) sulfate as a test for the presence of water
★ describe a physical test to show whether a sample of water is pure.

Reactions of metals with dilute acids

Magnesium, aluminium, zinc and iron all react smoothly with acids giving a salt and hydrogen.

$Mg(s)$ + $2HCl(aq)$ $\rightarrow$ $MgCl_2(aq)$ + $H_2(g)$
magnesium   hydrochloric acid   magnesium chloride   hydrogen

$Al(s)$ + $6HCl(aq)$ $\rightarrow$ $2AlCl_3(aq)$ + $3H_2(g)$
aluminium   hydrochloric acid   aluminium chloride   hydrogen

$Zn(s)$ + $H_2SO_4(aq)$ $\rightarrow$ $ZnSO_4(aq)$ + $H_2(g)$
zinc   sulfuric acid   zinc sulfate   hydrogen

$Fe(s)$ + $H_2SO_4(aq)$ $\rightarrow$ $FeSO_4(aq)$ + $H_2(g)$
iron   sulfuric acid   iron(II) sulfate   hydrogen

The combustion of hydrogen

Hydrogen/air mixtures are particularly dangerous for two reasons:

• they explode over a wide range of compositions
• their explosions are very violent.

A jet of pure hydrogen burns quietly with a hot, non-luminous flame (like a Bunsen burner with the air-hole fully open). If the flame is allowed to play on a cool surface the water formed in the combustion condenses to a liquid:

$$2H_2(g) + O_2(g) \rightarrow 2H_2O(l)$$

CAM

Hydrogen as a fuel

Hydrogen has been used as a fuel for rockets for decades, and once the engineering problems have been resolved, hydrogen vehicles are likely to enter commercial production fairly soon. Hydrogen has several advantages as a fuel:

• low density (the main reason for its use in rockets)
• the water formed during combustion can be condensed, so it will not add to global warming
• it can be obtained from water by electrolysis and, unlike oil, water is a renewable resource.

Tests for water

- The presence of water can be detected with white anhydrous $CuSO_4$, which turns blue as it reacts with water to form hydrated copper(II) sulfate (see Fig. 2e.01):

$$CuSO_4(s) + 5H_2O(l) \rightarrow CuSO_4.5H_2O(s)$$

Fig. 2e.01: White anhydrous CuSO₄ turns blue when water is added.

- Pure water boils at 100 °C and melts at 0 °C. Dissolved salts raise the boiling point and lower the freezing point of water.

The boiling point of water rises by about half a degree Celsius for every 30 g of sodium chloride dissolved in 1 dm^3 of water. A saturated solution of sodium chloride contains about 360 g of sodium chloride, which should increase the boiling point by $0.5 \times (360 \div 30) = 6$ °C. Brine (saturated sodium chloride solution) should boil at around 106 °C.

Use is made of the effect of dissolved salts on the lowering of the freezing point of water during cold weather. Rock salt (a mixture of grit and sodium chloride) is spread on icy roads in winter. As vehicles drive over the mixture, the pressure of their tyres on the grit causes some of the ice to melt. The water formed dissolves some of the salt. As the resulting solution has a lower freezing point than pure water it remains liquid. Eventually, as more and more vehicles use the road, all the ice is removed. As a guide, 230 g of sodium chloride in 1 dm^3 of water gives a solution that freezes at about –22 °C.

TIP

Note that the anhydrous copper(II) sulfate test only shows that water is **present**.

To **prove** that a liquid is water, it must turn anhydrous copper(II) sulfate blue **and** boil at 100 °C or melt at 0 °C.

CAM

Drinking water

Drinking water is obtained from rivers and reservoirs but water from these sources is not sufficiently pure to meet the stringent standards demanded of potable (drinking) water without further treatment. The water is first filtered to remove suspended solids; and then treated with chlorine to kill bacteria.

Uses of water

- As a solvent, e.g. in foodstuffs and many commercial products
- As a raw material, e.g. manufacture of sulfuric acid, ammonia and ethanol
- As a coolant, e.g. in power stations and in the manufacture of steel
- In the home, e.g. for washing and for flushing toilets.

You should now be able to:

★ describe the reactions of dilute hydrochloric acid and dilute sulfuric acid with some metals (see page 87)

★ describe the combustion of hydrogen (see page 87)

★ describe a test for the presence of water and a physical test to show whether or not a sample of water is pure (see page 88).

Practice questions

1. Write balanced equations, with state symbols, for the reaction between the following:

   (a) dilute hydrochloric acid and (i) zinc, (ii) aluminium **(6)**

   (b) dilute sulfuric acid and (i) magnesium, (ii) iron. **(6)**

2. How could you show that the liquid formed when hydrogen burns in air is pure water? **(3)**

3. (a) If hydrogen were to be used as a fuel, water would be the main product.

   From what you know about their physical properties, explain why it is easier to prevent atmospheric pollution by water than it is by carbon dioxide, which forms when fossil fuels are burned. **(3)**

   (b) Suggest two advantages and two disadvantages of hydrogen as a fuel for motor vehicles. **(4)**

F Reactivity series

You will be expected to:

★ describe how metals are arranged in a reactivity series based on their reactions and those of their compounds

★ describe how reactions with water and dilute acids can be used to deduce the order of reactivity of some metals

★ deduce the position of a metal in the reactivity series using reactions between metals, their oxides and aqueous solutions of their salts

★ describe oxidation and reduction as the addition and removal of oxygen respectively

★ define the terms *redox*, *oxidising agent* and *reducing agent*

★ identify the conditions under which iron will rust

★ describe how the rusting of iron may be prevented

★ explain the sacrificial protection of iron in terms of the reactivity series.

CAM

Properties and uses of metals

The properties of metals can make them very useful. Metals:

- are solids at room temperature (with the exception of mercury)

- have generally high melting points (e.g. iron 1535 °C, gold 1063 °C, copper 1083 °C, aluminium 660 °C), although there are exceptions (e.g. potassium 64 °C, sodium 98 °C)

- are shiny when freshly cut and, if they are not too reactive, they stay shiny (hence their use in mirrors and as decorative finishes, e.g. car bumpers)

- form alloys, which are mixtures of metals. Alloys have properties that differ from those of the metals from which they are made, which can make them more useful, e.g. brass (mixture of copper and zinc), solder (mixture of tin and lead), bronze (mixture of copper and tin)

- are usually strong – and can be hammered (malleable), bent or stretched (ductile) without breaking

- are good conductors of heat (hence their use in domestic heating systems and in car radiators)

- are good conductors of electricity (hence their use in electrical wiring).

In addition to the properties listed above, a group of metals known as the **transition metals** also:

- have high densities

- form coloured compounds

- act as catalysts, either as the pure metal or as compounds.

Common uses of some metals

copper

iron

aluminium

Fig. 2f.01: Common uses of metals

Copper	Iron	Aluminium	Zinc
water pipes (strong but easily bent to shape)	construction (as steel) (cheap and strong)	construction (where low density and strength essential, e.g. aircraft and high-voltage power cables)	galvanising iron and steel (see below)
electrical wiring (strong even when drawn into thin wires)	cooking utensils (good conductor of heat)	food containers (resistant to corrosion by acids in food)	a component of the alloy brass (see below)
cooking utensils (good conductor of heat)			

Steel is a mixture of iron, carbon and metallic elements, such as nickel and chromium; these mixtures are known as **alloys**. The composition of the steel limits the use that can be made of it.

Type of steel	Composition	Use
mild steel	0.2–0.4% carbon	car bodies, machinery
stainless steel	18% chromium, 8% nickel	chemical plant, cutlery

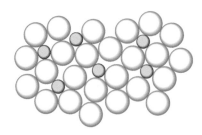

Fig. 2f.02: Diagram of an alloy – the small and large spheres represent atoms of different elements

The reactivity of metals

Metals differ greatly in their relative **reactivity**, for example rust quickly forms on iron in the presence of air and water, while gold is unaffected even after thousands of years.

We can arrange metals in a **reactivity series** from the most to the least reactive. This order is based on experiments to assess how vigorously metals react with:

- oxygen in the air
- water
- dilute acids
- aqueous solutions of salts of another metal.

 TIP The table shows the metals in the reactivity series that you need to remember, listed in order of decreasing reactivity. Learn their names, symbols and the order.

potassium	sodium	lithium	calcium	magnesium	aluminium	zinc	iron	copper	silver	gold
K	Na	Li	Ca	Mg	Al	Zn	Fe	Cu	Ag	Au

◀──────────────────── decreasing reactivity ────────────────────▶

Metals with oxygen

All metals (copper only superficially) react with oxygen in the air and can be made to burn in air if they are in the form of powder or as fine wire (e.g. steel wool).

This reaction is not a very accurate way of assessing the relative reactivities of K to Fe. Aluminium is also a problem (see page 95).

Metals with water

The pattern of relative reactivity is more obvious when metals are added to water.

Metal	Description of its reaction with water	Products of the reaction
K	violent: metal melts and the hydrogen formed catches fire	hydrogen + potassium hydroxide
Na	vigorous: metal melts but the hydrogen formed does not catch fire	hydrogen + sodium hydroxide
Li	fairly vigorous, but the metal does not melt	hydrogen + lithium hydroxide
Ca	brisk: a steady stream of hydrogen bubbles is given off	hydrogen + calcium hydroxide
Mg	no reaction with cold water reaction occurs if the magnesium is heated in steam	hydrogen + magnesium oxide
Zn	no reaction with cold water can be made to react with steam, but the metal must be heated strongly (for iron it must be red hot)	hydrogen + metal oxide
Fe		
Cu	no reaction	

Metals with dilute acids

The order of reactivity is confirmed when dilute acids are used.

- With sulfuric acid (H_2SO_4), a solution of the metal sulfate is formed.
- With hydrochloric acid (HCl), a solution of the metal chloride is formed.
- Hydrogen gas is given off with either acid.

TIP The reactions of potassium and sodium with dilute acids are too dangerous to carry out in a school laboratory.

Metal	Description of its reaction with acid	Products of the reaction
K Na	Explosive!	
Li	The metal reacts very vigorously and hydrogen is given off rapidly	hydrogen + solution of lithium sulfate or chloride (depending on the acid used)
Mg Ca Zn Fe	The metal dissolves and hydrogen is given off more and more slowly.	hydrogen + solution of metal sulfate or chloride (depending on the acid used)
Cu	No reaction	

Hydrogen in the reactivity series

Hydrogen can be placed (approximately) in the reactivity series from the reaction of metals with dilute hydrochloric acid; metals that react to form hydrogen are above it in the reactivity series, those that do not give hydrogen are below it in the reactivity series. On this basis hydrogen falls between iron and copper in the series.

The action of heat on metal nitrates and hydroxides

There is some correlation with the reactivity of a metal and the behaviour of its nitrate and hydroxide on heating, but one that is of little use in actually placing metals in the reactivity series. The results are summarised in the table below.

Metal	Action of heat on hydroxide	Action of heat on nitrate
Potassium and sodium	None	Decomposes to form the nitrite and oxygen $2MNO_3(s) \rightarrow 2MNO_2(s) + O_2(g)$
Lithium	Decomposes to form the oxide and water $2LiOH(s) \rightarrow Li_2O(s) + H_2O(l)$	Decomposes to form the oxide, nitrogen dioxide and oxygen $4LiNO_3(s) \rightarrow 2Li_2O(s) + 4NO_2(g) + O_2(g)$
Magnesium, calcium, zinc, copper	Decomposes to form the oxide and water $M(OH)_2(s) \rightarrow MO(s) + H_2O(l)$	$2M(NO_3)_2(s) \rightarrow 2MO(s) + 4NO_2(g) + O_2(g)$ [M = Mg, Ca, Zn or Cu]
Iron (III)	Decomposes to form iron(III) oxide and water $2Fe(OH)_3(s) \rightarrow Fe_2O_3(s) + 3H_2O(l)$	$2Fe(NO_3)_3(s) \rightarrow Fe_2O_3(s) + 6NO_2(g) + 1\frac{1}{2}O_2(g)$

The unusual behaviour of aluminium

Aluminium reacts vigorously with air, but the thin oxide layer that forms on the surface protects the metal underneath from further reaction. Aluminium therefore appears to be less reactive than it actually is unless special steps are taken to remove the oxide layer (for example with an abrasive) before carrying out any reactions.

Displacement reactions

If a metal is added to an aqueous solution (often the sulfate) of a second metal a reaction may or may not be observed: the outcome depends on the *relative* positions of the two metals in the reactivity series.

For example, the addition of a piece of zinc to an aqueous solution of copper sulfate leads to:

- gradual dissolving of the piece of zinc
- formation of a pink deposit of copper metal
- loss of the blue colour of the solution.

Zinc, the *more reactive* metal, has **displaced** copper ions from the solution. A **displacement reaction** has taken place.

Rule: If a reaction is observed when a metal (A) is added to an aqueous solution of a salt of a second metal (B) then metal A is more reactive than metal B.

Displacement reactions using metal salts

A reactivity series can be established by adding samples of metal to aqueous solutions of the salts of other metals and noting whether a reaction occurs. (K, Na and Ca are too reactive, and react with water in the solution.)

	magnesium sulfate solution	aluminium sulfate solution	zinc sulfate solution	iron sulfate solution	copper sulfate solution
magnesium		✓	✓	✓	✓
aluminium	✗		✓	✓	✓
zinc	✗	✗		✓	✓
iron	✗	✗	✗		✓
copper	✗	✗	✗	✗	

✓ = reaction observed   ✗ = no reaction

Displacement reactions using metal oxides

Metal oxides can be reduced by heating them with a more reactive metal. In practice it is not always easy to determine whether any reaction has in fact occurred because the metal formed by reduction looks very similar to that used as the reducing agent. The oxide and the metal should be finely powdered and well mixed to ensure that the reaction takes place as quickly and completely as possible.

The table below shows examples of reactions in which it is easy to see that a metal oxide has been reduced.

Reaction	Observation	Equation for reaction
copper(II) oxide + iron	black mixture turns pink as copper metal forms	$3CuO + 2Fe \rightarrow 3Cu + Fe_2O_3$
lead(IV) oxide + zinc	silvery beads of molten lead are formed	$PbO_2 + 2Zn \rightarrow Pb + 2ZnO$
iron(III) oxide + magnesium	exothermic reaction, and iron formed is magnetic	$Fe_2O_3 + 3Mg \rightarrow 2Fe + 3MgO$

Using reactions of this kind, it is possible to establish a reactivity series broadly similar to that obtained from the reactions between metals and their salts in aqueous solution.

Chemistry A Study Guide*

Reduction of metal oxides with carbon

The ability, or otherwise, of carbon to reduce the oxide of a metal provides another way of placing it in the reactivity series:

- Not reduced on heating with carbon – the oxides of potassium, sodium, lithium, magnesium and calcium.
- Reduced on heating with carbon – the oxides of zinc, iron and copper.

Oxidation and reduction

- The addition of oxygen to a substance during a chemical reaction is known as **oxidation**, e.g.

    $C(s) + O_2(g) \rightarrow CO_2(g)$ and $2Mg(s) + O_2(g) \rightarrow 2MgO(s)$

    The oxygen involved does not have to be oxygen gas. In the reaction

    $CuO(s) + Mg(s) \rightarrow Cu(s) + MgO(s)$

    the magnesium has been oxidised by gaining oxygen from the copper(II) oxide.

- The substance that supplies the oxygen in the reaction is known as an **oxidising agent**, common examples being oxygen itself or the oxide of a less reactive metal (as shown above).

 TIP The addition of oxygen to the metals described above is called oxidation. However, the reaction of methane with oxygen is more usually described as **combustion**, although it involves oxidation of the methane.

- The removal of oxygen from a substance is known as **reduction**.

    | $2PbO(s)$ | + | $C(s)$ | $\rightarrow$ | $2Pb(s)$ | + | $CO_2(g)$ |
    | lead(II) oxide | + | carbon | $\rightarrow$ | lead | + | carbon dioxide |
    | $CuO(s)$ | + | $H_2(g)$ | $\rightarrow$ | $Cu(s)$ | + | $H_2O(l)$ |
    | copper oxide | + | hydrogen | $\rightarrow$ | copper | + | water |

- The oxygen is removed by substances known as **reducing agents**, common examples being carbon and hydrogen.

- Note that when oxidation occurs in a reaction, reduction also occurs, e.g.:

    $CuO(s) + Mg(s) \rightarrow Cu(s) + MgO(s)$

    The magnesium has been oxidised by gaining oxygen from the copper(II) oxide, while the CuO has lost oxygen and has been reduced. The oxidation and reduction together constitute a **redox** reaction.

Redox in terms of electron transfer

In some reactions involving oxidation and reduction (redox) no oxygen is involved. For example, if zinc powder is added to aqueous copper(II) sulfate the following observations can be made

- the blue solution of copper(II) sulfate becomes paler
- the zinc reacts and disappears
- metallic copper is formed.

The equation for the reaction is $Zn(s) + CuSO_4(aq) \rightarrow Cu(s) + ZnSO_4(aq)$

The zinc metal has lost electrons and has been converted to zinc ions: $Zn(s) - 2e^- \rightarrow Zn^{2+}(aq)$

Loss of electrons is defined as **oxidation**.

The copper ions have gained electrons and have been converted to copper metal: $Cu^{2+}(aq) + 2e^- \rightarrow Cu(s)$

Gain of electrons is defined as **reduction**.

Remember: OILRIG – Oxidation Is Loss of electrons, Reduction Is Gain of electrons.

When redox reactions take place the **oxidation state** of the substances involved changes. Using the example above:

Reactant	Product	Initial oxidation state	Final oxidation state	Change in oxidation state
Zn	Zn^{2+} / Zn(II)	0	+2	Increase – the Zn has been oxidized
Cu^{2+} / Cu(II)	Cu	+2	0	Decrease – the Cu^{2+} has been reduced

The oxidation state is shown as a Roman numeral after the symbol for the ion, for example Cu(II) is 'copper two', 'iron three' would be written as Fe(III). By definition all metals have an oxidation state of zero.

Remember: If a species is oxidised its oxidation number increases, if it is reduced its oxidation number decreases.

Colour changes can also be used to show that a redox reaction has taken place. Aqueous potassium manganate(VII) behaves as an oxidising agent in acidic solution, while potassium iodide is a reducing agent.

Substance	Initial colour of solution	Final colour of solution	Chemical change
Potassium manganate(VII) [oxidising agent]	deep purple	colourless	manganate(VII) $\rightarrow$ manganese(II) Oxidation state decreases from VII to II
Potassium iodide [reducing agent]	colourless	red-brown	iodide $\rightarrow$ iodine Oxidation state increases from −1 to 0

The rusting of iron

Iron is not very reactive but, if left exposed to air and water for prolonged periods, it forms hydrated iron(III) oxide, commonly known as **rust**. This is unsightly and reduces the strength of the metal.

Air and water must *both* be present for rusting to occur, as shown by placing an iron nail in tubes containing:

- water open to the air (tube contains water and air) – nail rusts
- air and a drying agent (tube contains air but no water) in sealed tube – nail does not rust.
- boiled water (tube contains water but no air) in sealed tube – nail does not rust

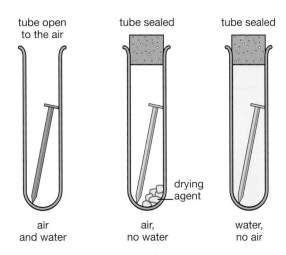

2f.03: Testing iron nails for rusting

Prevention of rusting

The rate of rusting of iron can be reduced by protecting its surface to exclude air and moisture with:

- paint, such as on a car – relatively cheap and easy to cover large surfaces
- grease or oil, such as on a cycle chain – easy to reapply on surfaces that are continually rubbed (also acts as lubrication)
- plastic – can be coloured, as in green garden fencing wire.

However, if the coating is broken, rusting begins again.

Rusting can also be prevented by dipping the iron article into molten zinc (which has a relatively low melting point of 420 °C) to coat the iron. This process is known as **galvanising**. This is better than coating the surface because, even if the zinc coating is scratched, the iron will not rust because the zinc continues to react preferentially with the air and water.

Sacrificial protection

Large steel structures (e.g. oil rigs, pipelines and ships) are too large to be galvanised. Instead plates of zinc or magnesium alloy are bolted to their surface. Zinc and magnesium are higher up the reactivity series than iron, so they react with air and moisture in preference to the iron or steel. The plates are sacrificed to protect the iron or steel, hence the name **sacrificial protection**. The plates have to be replaced periodically as they corrode.

Fig. 2f.04: Sacrificial protection of ship hull using zinc blocks

CAM

The reactivity series and simple cells

If two copper electrodes in a beaker of dilute sulfuric acid are connected by a voltmeter, no reading is obtained because the electrodes have the same reactivity. However, if one of the copper electrodes is replaced by a zinc electrode the voltmeter shows a reading – electrical energy has been produced.

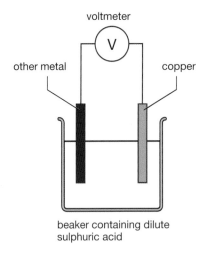

Fig. 2f.05: A simple cell

This electrical energy takes the form of a flow of electrons around the circuit. A **simple cell** has been created, and it is found that the copper electrode is the positive terminal of the cell. Electrons flow from the more reactive metal (zinc) through the external circuit to the less reactive metal (copper).

If the zinc electrode is replaced in turn by other metals (keeping one electrode as copper) the voltage generated is found to depend on the metal used:

Metal	Magnesium	Calcium	Zinc	Iron
Voltage	2.72	3.21	1.00	0.78

These results show that the greater the difference in the reactivity of the two electrodes used in the cell, the larger the voltage produced by the cell, although magnesium and calcium do not fit the pattern because the voltage depends on other factors that are not linked to reactivity in a straightforward way.

Differences in reactivity between two metals formed the basis for the construction of the earliest cells, but even the most modern of batteries depends on the same principles. All batteries contain a substance that is oxidised (gives up electrons) and a substance that is reduced (takes up electrons) and the electrons are transferred through the external circuit (wires). A motor or bulb in the external circuit converts the chemical energy of the redox reaction to motion or light, respectively.

You should now be able to:

★ define the term *reactivity series* (see page 93)

★ describe experiments to arrange the following metals in increasing order of reactivity: potassium, sodium, lithium, calcium, magnesium, aluminium, zinc, iron and copper (see page 93)

★ define the terms *oxidation*, *reduction*, *redox*, *oxidising agent* and *reducing agent* (see page 97)

★ describe the conditions under which iron will rust (see page 99)

★ describe a range of methods used to prevent the rusting of iron (see page 99)

★ explain how sacrificial protection works in terms of the reactivity series (see page 100).

Practice questions

1. A new metal has been discovered and given the name 'millennium'. It is shiny and conducts heat well. Millennium reacts with sulfuric acid to produce a steady stream of gas bubbles. If a strip of millennium is placed in copper sulfate solution a reaction takes place. A piece of magnesium ribbon placed in millennium sulfate solution soon becomes covered with shiny crystals. Millennium burns brightly if heated in air, to give a white powder.

   (a) Apart from being shiny and conducting heat well, state three other properties that you would expect millennium to have. **(3)**

   (b) What *three things* would you expect to see when millennium was added to copper sulfate solution? **(3)**

   (c) (i) Name the gas formed when millennium reacts with sulfuric acid. **(1)**

       (ii) Write a word equation for the reaction in part (i). **(2)**

   (d) (i) What are the shiny crystals formed on the surface of the magnesium? **(1)**

       (ii) Write a word equation for the formation of the crystals. **(3)**

   (e) (i) Place the elements copper, millennium and magnesium in order of increasing reactivity, starting with the least reactive. **(2)**

       (ii) Explain briefly how you arrived at this order of reactivity. **(3)**

   (f) Name the white powder formed when millennium burns in air. **(1)**

2. (a) (i) Carbon is a reducing agent. Explain what is meant by the term reducing agent. **(2)**

       (ii) Lead oxide is a yellow solid. What would you expect to see if a mixture of powdered lead oxide and carbon was heated? **(2)**

       (iii) Write a symbol equation for the reaction that takes place. **(2)**

3. Choose the most appropriate method for protecting each of the following from rusting. In each case, explain your choice.

   (a) steel dustbin, (b) steel bicycle chain, (c) steel garden spade, (d) iron railings,
   (e) iron girders on an oil rig **(10)**

4. Strips of several metals of equal mass were added separately to excess dilute hydrochloric acid in the apparatus shown below. The time taken for 10 cm³ of hydrogen to be collected was measured. The results are shown in the table.

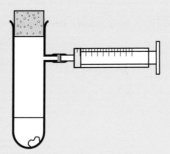

Metal	Time taken (seconds)
calcium	10
iron	30
magnesium	12
zinc	25

(a) (i) On the evidence provided by these results, state and explain which metal is least reactive. **(2)**

   (ii) State and explain what other factor has to be controlled to allow a fair comparison of the reactivity of the four metals. **(2)**

(b) Once the experiment had been modified to ensure that it was a fair test, a strip of aluminium produced very little hydrogen in the hydrochloric acid unless the surface was first rubbed with emery paper. Aluminium is more reactive than zinc. Suggest an explanation for this unexpected result. **(3)**

(c) Explain why painting iron articles helps to prevent rusting. **(2)**

(d) Galvanising involves coating iron articles with zinc. Use information from the table to help you to explain why galvanising iron protects it from rusting more effectively than painting does. **(5)**

5. The usual way of placing metals in a reactivity series is to add one metal to an aqueous solution of the sulfate of a second metal and to see whether or not a displacement reaction occurs.

Fred's teacher told him that another way of placing metals in a reactivity series was to measure the voltage set up between two different metals placed in a beaker of sulfuric acid. A suitable apparatus is shown in the diagram.

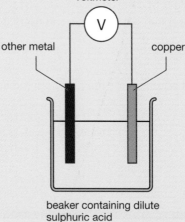

voltmeter

other metal   copper

beaker containing dilute
sulphuric acid

(a) In some experiments in which copper was always used as one of the metals, the following results were obtained.

Other metal	Reading on voltmeter (volts)
aluminium	2.00
iron	0.80
magnesium	2.70
zinc	1.10

Describe how these voltages are related to the position of the metals in the reactivity series shown below, which was obtained from displacement reactions. **(2)**

most reactive	magnesium	aluminium	zinc	iron	copper	least reactive

(b) Iron can be prevented from rusting by fixing a piece of magnesium metal to it. Explain how this prevents the iron from rusting. **(3)**

(c) Fred did more experiments with two other metals, nickel and tin. Fred's results for four metals he studied are shown on the table below.

First metal	Other metal	Voltage (volts)
copper	iron	0.80
copper	nickel	0.60
copper	tin	0.50
copper	zinc	1.10

(i) Put iron, nickel, tin and zinc into a reactivity series, starting with the most reactive **(1)**

(ii) Explain how you arrived at the reactivity order given in part (c)(i). **(3)**

(d) Baked bean cans are made of iron and have a coating of tin on them.

Use the information from part (c) to explain why the iron can rusts if the coating of tin is scratched. **(3)**

(e) (i) State and explain what would you expect to see if nickel powder was added to a solution of copper sulfate. **(4)**

(ii) Copy and complete the following equation for the reaction, including state symbols.

$Ni(s) + Cu^{2+}(aq) \rightarrow$ **(2)**

G Tests for ions and gases

You will be expected to:

* ★ describe simple tests for some cations
* ★ describe simple tests for some anions
* ★ describe simple tests for some gases.

Tests for cations

Using flame tests

A sample of a salt of a Group 1 or 2 metal often imparts a characteristic colour to a Bunsen flame (see Fig. 2g.01). A small quantity of the solid salt is introduced to the flame on a platinum wire, which has previously been cleaned by dipping it into concentrated hydrochloric acid and then heating to red heat.

Metal	Flame colour
lithium / Li^+ ions	crimson red
sodium / Na^+ ions	golden yellow
potassium / K^+ ions	lilac
calcium / Ca^{2+} ions	brick red

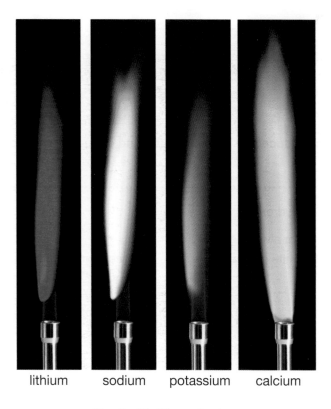

lithium   sodium   potassium   calcium

Fig. 2g.01: Flame tests

Using sodium hydroxide solution

Ion	Test reagent	Observation if ion is present
copper(II) / Cu^{2+} iron(II) / Fe^{2+} iron(III) / Fe^{3+} aluminium / Al^{3+} zinc (II) / Zn^{2+}	aqueous sodium hydroxide	blue precipitate green precipitate red-brown precipitate white precipitate, soluble in excess sodium hydroxide solution white precipitate, soluble in excess sodium hydroxide solution
ammonium / NH_4^+	warm with aqueous sodium hydroxide	ammonia gas is formed, which turns moist red litmus paper blue

CAM

Fig. 2g.02: Precipitates produced by iron(II), iron(III) and copper(II) with sodium hydroxide solution

CAM

Using aqueous ammonium hydroxide

- Aluminium / Al^{3+} ions form a white precipitate.
- Zinc / Zn^{2+} ions form a white precipitate, soluble in excess ammonium hydroxide solution.

Tests for anions

Ion	Test reagent	Observation if ion present
chloride ions / Cl^- bromide ions / Br^- iodide ions / I^-	aqueous silver nitrate acidified with dilute nitric acid	white precipitate pale yellow precipitate yellow precipitate
sulfate / SO_4^{2-}	aqueous barium chloride acidified with dilute hydrochloric acid	white precipitate
carbonate ions / CO_3^{2-}	add dilute acid	carbon dioxide formed, turns lime water milky
nitrate ions / NO_3^-	heat with aluminium and aqueous sodium hydroxide solution	ammonia formed, turns moist red litmus paper blue

Tests for gases

Gas	Test
hydrogen	burns with a squeaky pop when ignited
oxygen	ignites a glowing splint
carbon dioxide	turns lime water [$Ca(OH)_2(aq)$] milky
ammonia	turns moist red litmus paper blue (NH_3 is alkaline)
chlorine	turns moist blue litmus paper red, then bleaches it

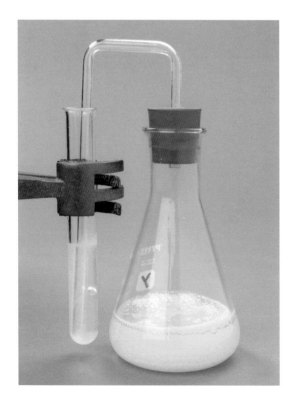

Fig. 2g.03: Testing for carbon dioxide and oxygen

You should now be able to recall how to test for each of the following in the laboratory (see pages 105, 106, 107).

Cations	Anions	Gases
lithium	chloride	hydrogen
sodium	bromide	oxygen
potassium	iodide	carbon dioxide
calcium	sulfate	ammonia
ammonium	carbonate	chlorine
copper(II)		
iron(II)		
iron(III)		

Practice questions

1. Copy the table and complete the blanks.

Ion tested for	Test reagent	Observation
calcium / Ca^{2+}	_____ **(1)**	_____ _____ **(2)**
copper(II) / Cu^{2+} iron(II) / Fe^{2+} iron(III) / Fe^{3+}	aqueous sodium hydroxide	_____ precipitate _____ precipitate _____ precipitate **(3)**
sulfate / SO_4^{2-}	_____ _____ **(2)**	_____ precipitate **(1)**
chloride ions / Cl^- bromide ions / Br^- iodide ions / I^-	_____ _____ **(2)**	_____ precipitate _____ precipitate _____ precipitate **(3)**

2. The following pairs of solutions are mixed and thoroughly shaken. Name any precipitate remaining after this has been done. If there is no precipitate, write 'no precipitate'.

   (a) sodium chloride and silver nitrate in dilute nitric acid

   (b) sodium carbonate and dilute hydrochloric acid

   (c) zinc sulfate and excess ammonium hydroxide

   (d) copper sulfate and sodium hydroxide. **(4)**

Section Three

3 Organic chemistry

A Alkanes

> **You will be expected to:**
>
> ★ explain the terms *homologous series*, *hydrocarbon*, *saturated*, *unsaturated*, *general formula* and *isomerism*
> ★ recall and use the general formula for alkanes: C_nH_{2n+2}
> ★ draw displayed formulae for some simple alkenes
> ★ name the straight-chain isomers of some simple alkenes
> ★ name the products of the complete and incomplete combustion of alkanes
> ★ describe the reaction of methane with bromine and chlorine in the presence of UV light.

Organic chemistry introduction

Organic chemistry is the chemistry of the element carbon. Carbon atoms have the unique property of joining to other carbon atoms to form long chains of atoms. Carbon on its own forms more compounds than all of the 90 or so remaining elements in the Periodic Table do together.

- Compounds that contain only carbon and hydrogen atoms are called **hydrocarbons**.
- The hydrocarbons are divided into groups of compounds with the same general formula and similar chemical properties. Each group of compounds is called a **homologous series**.
- Examples of homologous series are the alkanes and alkenes.

The structure of alkanes

- The homologous series of alkanes have formulae that can all be represented by the **general formula** C_nH_{2n+2}, where n is a whole number. For example, when $n = 5$ we obtain $C_5H_{(2 \times 5) + 2} = C_5H_{12}$
- The names and formula of the first five alkanes are shown in the table below.

Name	Formula
methane	CH_4
ethane	C_2H_6
propane	C_3H_8
butane	C_4H_{10}
pentane	C_5H_{12}

- Alkanes are **saturated hydrocarbons** because they contain only *single* C–C and C–H bonds. In alkanes each carbon atom shares an electron with *four* other atoms.
- The '-ane' ending to the name shows that it refers to an alkane.
- The first part of the name tells you how many carbon atoms there are in the longest *unbranched* chain (see below).

Name	Number of carbon atoms in the longest *unbranched* chain
meth-	1
eth-	2
prop-	3
but-	4
pent-	5

- An extra CH_2 unit is added for each additional carbon added to the chain.

Displayed formulae

Displayed formulae try to show the arrangement of the atoms in space.

Here are the two-dimensional displayed formulae for the first five alkanes.

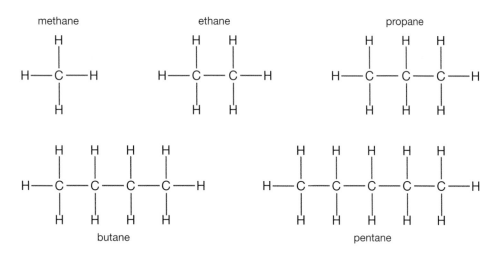

Fig. 3a.01: Two-dimensional displayed formulae of alkanes

In reality, the bonds on each carbon are directed towards the corners of a *regular tetrahedron*, and the bond angle is 109°, not 90° as shown in the two-dimensional diagrams in Fig. 3a.01.

Representation of the three-dimensional shapes of methane and pentane are shown in Fig. 3a.02.

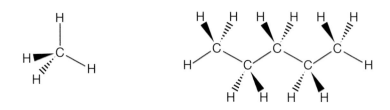

Fig. 3a.02: Three-dimensional displayed formulae of alkanes

Isomers

When there are three or more carbon atoms in a chain, a further carbon atom can be added either on the *end of* or *in the middle of* the carbon chain.

For example:

$$CH_3 - CH_2 - CH_2 - CH_3$$

$$CH_3 - CH - CH_3$$
$$|$$
$$CH_3$$

Fig. 3a.03: CH_3 added to the end of a C_3 alkane chain

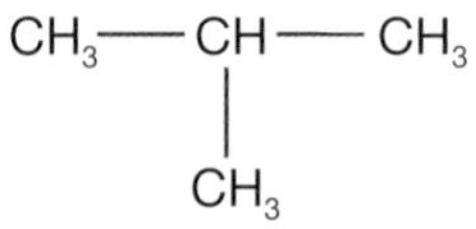

Fig. 3a.04: CH_3 added to the middle of a C_3 alkane chain

The resulting structures are known as **isomers**. Isomers have:

- the same **molecular formula** (in this case C_4H_{10})
- different **structural formulae**.

Isomers without any branches along the chain are **straight-chain isomers**, e.g. butane, pentane.

Number of C atoms	Straight-chain isomer	Branched isomers	

4	butane
	$CH_3 - CH_2 - CH_2 - CH_3$
		$CH_3 - CH - CH_3$
		      $\|$
		     CH_3	

5	pentane
	$CH_3 - CH_2 - CH_2 - CH_2 - CH_3$
		$CH_3 - CH - CH_2 - CH_3$
		       $\|$
		      CH_3
			CH_3
			 $\|$
			$CH_3 - C - CH_3$
			 $\|$
			CH_3

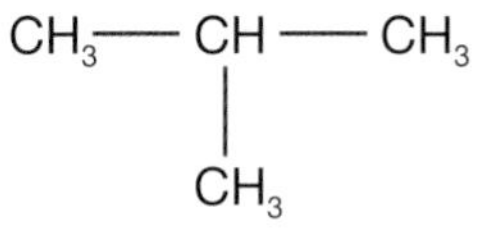

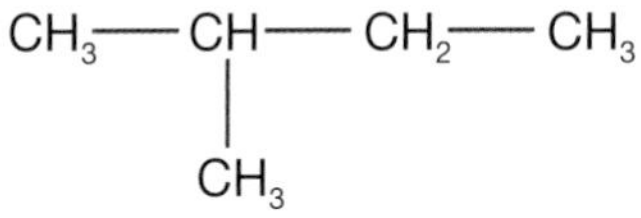

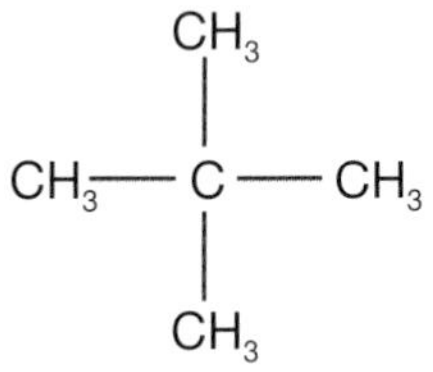

> **TIP**
>
> You only have to remember the names of straight-chain isomers.
>
> Note: butane and the 4-carbon branching isomer are isomers of the alkane with the formula C_4H_{10} **not** isomers of each other. Even experienced chemists can get this wrong and talk about 'isomers of butane'. A hydrocarbon with a name, like butane or hexane, cannot have isomers. The structure in the top right of the table is called methyl propane.

Combustion of alkanes

The combustion of alkanes can be represented (using methane as an example) as follows:

- in a plentiful supply of air /excess oxygen, **complete combustion**: $CH_4 + 2O_2 \rightarrow CO_2 + 2H_2O$
- in a restricted supply of air / limited oxygen, **incomplete combustion**: $CH_4 + 1\frac{1}{2}O_2 \rightarrow CO + 2H_2O$

Incomplete combustion leads to the formation of poisonous (toxic) carbon monoxide and is dangerous if it occurs in confined spaces (e.g. in buildings as a result of faulty gas boilers etc.).

Reaction of alkanes with halogens

In ultra-violet (UV) light (such as in sunlight), the halogen bromine reacts with methane to produce bromomethane. Hydrogen bromide gas is the other product. The UV light provides the energy needed to break the bond in the halogen molecules to start the reaction.

$$CH_4 + Br_2 \rightarrow CH_3Br + HBr$$
$$\text{bromomethane}$$

Note, if the reaction continues, a mixture of halogenated products is formed. This is not a useful synthesis of halogenoalkanes because separation of the desired product is tedious.

You should now be able to:

★ explain the meaning of the terms *homologous series*, *hydrocarbon*, *saturated*, *general formula* and *isomerism* (see pages 112, 113, 114)

★ state the general formula of the alkanes (see page 112)

★ draw the displayed formulae for alkenes with up to 5 carbon atoms and name the straight-chain isomers (see page 113)

★ name the products of complete and incomplete combustion of alkanes (see page 115)

★ name the reaction conditions required, and the products, when methane reacts with bromine (see page 115).

Practice questions

1. a) Explain the meaning of the term *isomer*   (2)

    (b) Draw the isomers of the alkane C_4H_{10} Name the straight-chain isomer.   (3)

2. Write word equations for the burning of propane in the following conditions:

    (a) in a plentiful supply of oxygen

    (b) in a restricted supply of oxygen.   (4)

3. (a) Give the general formula for an alkane with *n* carbon atoms.   (1)

    (b) Give the formula of the alkane with five carbon atoms.   (1)

4. (a) Under what conditions will methane react with bromine?   (1)

    (b) Write a balanced equation for the reaction to form bromomethane.   (2)

B Alkenes

You will be expected to:

★ state that alkenes have the general formula C_nH_{2n}
★ draw displayed formulae for some simple alkenes
★ name the straight-chain isomers of some simple alkenes
★ describe how alkenes undergo an addition reaction with bromine
★ explain that the addition reaction with bromine is a test for alkenes.

The structure of alkenes

- The alkenes form a homologous series with the general formula C_nH_{2n}
- The first three hydrocarbons in the alkene series are:
  ethene C_2H_4
  propene C_3H_6
  butene C_4H_8
- The '-ene' ending to the name shows that it refers to an alkene.
- Alkenes contain a C=C **double bond** in addition to C–C and C–H bonds. The carbon atoms of the C=C bond share electrons with only *three* other atoms.
- An extra CH_2 unit is added for each additional carbon added to the chain.
- Alkenes are **unsaturated hydrocarbons** because the C=C bond allows other atoms to bond with the C=C carbons without the need to break the C=C bond completely.

Displayed formulae of alkenes

The bond angles around the C=C bond are 120°. So, in alkenes, the C=C bond and the four atoms or groups bonded to it lie in the same plane.

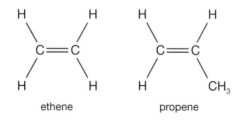

ethene   propene

Fig. 3b.01: Ethene and propene

If the alkene contains more than three carbon atoms, the C=C can be in different places:

- but-1-ene **CH$_2$=CH**–CH$_2$–CH$_3$: the double bond starts on the '1' carbon, the first in the chain
- but-2-ene CH$_3$–**CH=CH**–CH$_3$: the double bond starts on the '2' carbon, the second in the chain.

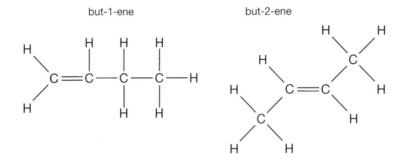

Fig. 3b.02: But-1-ene and but-2-ene

But-1-ene and but-2-ene are **positional isomers** of C_4H_8 – the C=C is in different positions in the two molecules.

Addition reactions of alkenes

Small molecules, such as bromine, water and hydrogen can add across the double bond in the unsaturated alkenes because they can form additional bonds without fully breaking any other bonds first. The original alkene survives intact plus the added small molecule. (By contrast, alkanes are said to be 'saturated' because they cannot form any further bonds without *first* breaking some existing C–C or C–H bonds.)

If an orange aqueous solution of bromine (bromine water) is shaken with an alkene the orange colour fades to colourless.

$$CH_3\text{–}CH=CH_2 \; + \; Br_2 \qquad \rightarrow \qquad CH_3\text{–}CHBr\text{–}CH_2Br$$

propene   bromine(orange)   1,2-dibromopropane (colourless)

Test for an alkene

The disappearance (discharge) of the colour of the bromine water is a *test* for the presence of the C=C bond. A positive result distinguishes alkenes from alkanes, which give a negative result.

You should now be able to:

★ state the general formula of the alkenes (see page 117)

★ describe the addition reaction of ethene with bromine (see page 118)

★ describe the test that distinguishes between an alkane and alkene, and explain the differing results (see page 118).

Practice questions

1. (a) (i) Write down the general formula of the alkenes. **(1)**

   (ii) Draw the displayed (full structural) formulae of the straight-chain isomers of the alkene containing four carbon atoms. Name each isomer. **(4)**

   (ii) Draw the displayed (full structural) formulae of the branched-chain isomer of the alkene containing four carbon atoms. **(2)**

2. (a) Describe what you would do in the laboratory to distinguish a liquid alkene from a liquid alkane, e.g. hexene and hexane. **(2)**

   (b) Describe what you would observe *with both liquids* during the test. **(3)**

C Ethanol

You will be expected to:

★ describe the manufacture of ethanol from ethene and steam
★ describe the manufacture of ethanol from the fermentation of sugars
★ evaluate factors for choosing which method to use for the manufacture of ethanol
★ describe how ethanol can be dehydrated to form ethene.

CAM ★ recall that ethanol is used as a solvent and as a fuel.

Manufacture of ethanol

Ethanol is an alcohol that can be manufactured in different ways.

Ethanol from ethene

Ethene is reacted with steam at 70 atmospheres pressure and 300 °C in the presence of a phosphoric acid catalyst.

The process is **continuous**, which means that there is no break in the production of ethanol once the ethene and steam have been brought together over the catalyst.

Ethanol from sugars

Any plant material containing sugar can produce ethanol by fermentation in the presence of yeast. Sugar cane and maize are often used because they can be grown efficiently, particularly in warm countries, which have long growing seasons.

An aqueous solution of the sugar and yeast are mixed and kept at a temperature of around 30 °C until fermentation ceases, usually when the percentage of ethanol in the mixture is about 12%. The action of the yeast is inhibited by concentrations of ethanol higher than this. Carbon dioxide is also formed during the fermentation process.

If more ethanol is wanted, another fermentation must be started – the process of ethanol production is not continuous, but is a **batch process**. This is in contrast to the continuous process producing ethanol from ethene.

Comparison of methods

The different processes for manufacturing ethanol have different advantages and disadvantages.

	Advantages	Disadvantages
Fermentation from sugars	• made from a renewable resource (plants)	• process is slow • a batch process • resulting ethanol/water mix must pass through fractional distillation to produce pure ethanol
From ethene	• process is faster • process is continuous • pure ethanol is obtained, so no need for fractional distillation	• ethene is obtained from crude oil, a **non-renewable resource**

Reactions of ethanol

Dehydration

Dehydration is the removal of water. Ethanol is **dehydrated** to form ethene by passing ethanol vapour over heated aluminium oxide.

The equation for the reaction is:

$$C_2H_5OH \rightarrow C_2H_4 + H_2O$$

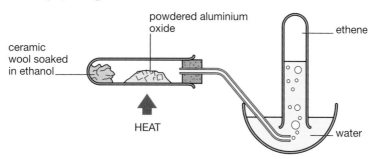

Fig. 3c.01: Dehydration of ethanol to ethene

CAM

Combustion

Alcohols burn with a hot, smoke-free flame. The following equation describes the combustion of ethanol.

$$C_2H_5OH + 3O_2 \rightarrow 2CO_2 + 3H_2O$$

Uses of ethanol

Ethanol, dyed purple and mixed with methanol (CH_3OH) to give it a disgusting taste, is sold as methylated spirit. It burns with a hot, clear blue, smokeless flame and is widely used as a convenient liquid fuel.

In some developing countries (notably in South America) the relatively cheap ethanol is added to petrol to give 'gasohol' which reduces the need to import expensive petrol.

Ethanol is also widely used in the manufacture of flavourings (esters), in the cosmetics industry and as a solvent.

You should now be able to:

★ describe two methods for manufacturing ethanol (see page 120)
★ evaluate the advantages and disadvantages of the different ways of manufacturing ethanol (see page 120)
★ describe how ethanol can be dehydrated and name the product formed (see page 121).
CAM ★ describe two uses of ethanol (see page 121).

Practice questions

1. (a) (i) Write a word *equation* that describes the process of fermentation to manufacture ethanol. **(4)**

    (ii) Why does fermentation cease naturally? **(1)**

    (iii) By what process is pure ethanol obtained following fermentation? **(2)**

    (iv) On what physical property does this process depend? **(1)**

(b) (i) Give the reagents and conditions for the manufacture of ethanol from ethene. **(4)**

    (ii) Suggest two advantages of the method described in (b)(i) over fermentation for the production of ethanol. **(2)**

    (iii) In some countries ethanol is manufactured by fermentation. What use is made of the ethanol and why is economical? **(2)**

(c) (i) Describe two features of the combustion of ethanol that make it a useful fuel. **(2)**

    (ii) Copy and balance the equation below for the complete combustion of ethanol. **(2)**

$$C_2H_5OH \ + \ \underline{\hspace{2cm}} \ O_2 \rightarrow \underline{\hspace{2cm}} \ CO_2 \ + \ \underline{\hspace{2cm}} \ H_2O$$

    (iii) Give one other use of ethanol in manufacturing, other than as a fuel. **(1)**

2. (a) Label the diagram below, which involves the formation of ethene from ethanol. **(3)**

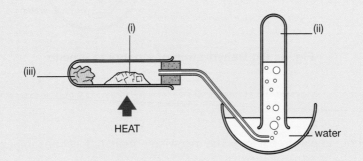

(b) Name the type of reaction taking place in this experiment. **(1)**

(c) How could you show that the gas collected in this experiment contained a C=C bond? **(3)**

D Carboxylic acids

You will be expected to:

★ state the general formula of carboxylic acids and recall the functional group
★ name and draw the structure of simple carboxylic acids
★ describe how carboxylic acids can be made by oxidising alcohols
★ explain why carboxylic acids are known as weak acids
★ describe how carboxylic acids react with alcohols to form esters.

Homologous series

The carboxylic acids are an example of the **homologous series** that has the general formula $C_nH_{2n+1}COOH$ (where n is a whole number).

- methanoic acid   HCOOH
- ethanoic acid CH_3COOH
- propanoic acid C_2H_5COOH
- butanoic acid C_3H_7COOH

The –COOH part of the molecule is known as the **functional group**. It is this group that gives the carboxylic acids their identical characteristic reactions.

Structural formulae

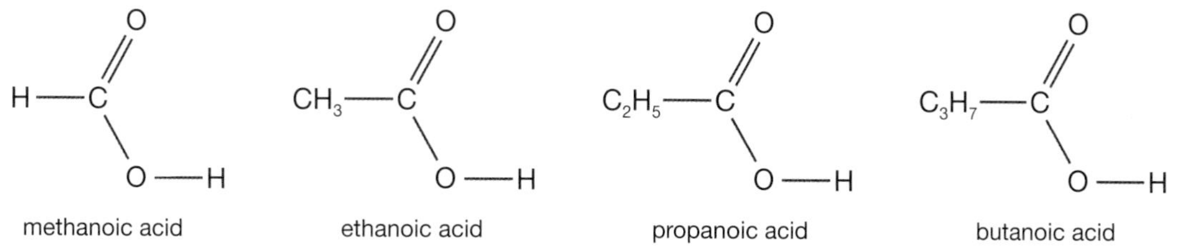

methanoic acid   ethanoic acid   propanoic acid   butanoic acid

Fig 3d.01: Structural formulae of carboxylic acids

Formation of ethanoic acid

Ethanol can be oxidised to ethanoic acid in two ways: biologically and chemically.

Biological oxidation

Under anaerobic conditions sugars undergo fermentation to produce a dilute aqueous solution of ethanol. If this solution is allowed to ferment further under partially aerobic conditions (the oxygen level must be carefully controlled) and in the presence of certain bacteria, a solution of ethanoic acid is produced. This reaction is important commercially in the manufacture of wine vinegar.

Section 3D Carboxylic acids

Chemical oxidation

Heating ethanol with acidified potassium manganate(VII) oxidises it to ethanoic acid. The equation for the reaction is complicated, but can be more simply represented as follows:

$C_2H_5OH + 2[O] \rightarrow CH_3COOH + H_2O$, where [O] represents the oxygen from the oxidising agent.

During the reaction the intense purple colour of the oxidising agent is lost as it is converted to the colourless manganese(II) ion.

Weak acids

Carboxylic acids are acids because they **dissociate** in water to give hydrogen ions and ethanoate ions:

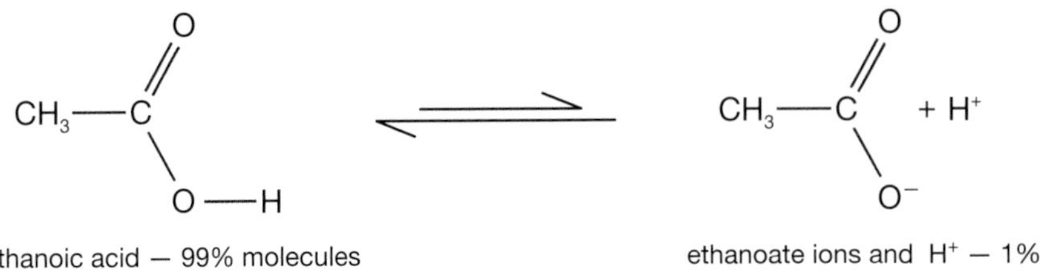

ethanoic acid — 99% molecules   ethanoate ions and H^+ — 1%

Fig 3d.02: Dissociation of ethanoic acid

Carboxylic acids are **weak acids**. This means they produce fewer hydrogen ions in water than the **strong acids**, such as hydrochloric acid, nitric acid and sulfuric acid.

In a solution of ethanoic acid in water, only about 1% of the ethanoic acid molecules dissociate, while in a solution of hydrochloric acid in water, *all* the HCl molecules dissociate.

The pH of hydrochloric acid would be *lower* than that of a solution of ethanoic acid of the same concentration.

The formation of esters

Ethanol, C_2H_5OH, reacts on heating with ethanoic acid, CH_3COOH, in the presence of a little concentrated sulfuric acid catalyst to form ethylethanoate and water:

$$C_2H_5OH + CH_3COOH \rightarrow CH_3COOC_2H_5 + H_2O$$

Ethylethanoate is an example of an **ester** and the reaction above occurs with any combination of carboxylic acid and alcohol. Esters have characteristically sweet odours and many fruit flavours are due to esters. Large quantities are manufactured commercially for artificial food flavourings.

> **You should now be able to:**
>
> ★ state the general formula of carboxylic acids (see page 123)
> ★ state the functional group in carboxylic acids (see page 123)
> ★ describe how carboxylic acids can be prepared (see pages 123, 124)
> ★ explain the meaning of the term **weak** acid (see page 124)
> ★ describe how to make an ester from a carboxylic acid (see page 124).

Practice questions

1. Draw the structural formulae of the following:

   (a) methanoic acid, (b) ethanoic acid, (c) propanoic acid. **(3)**

2. (a) Ethanoic acid is a weak acid. What is meant by the term 'weak acid'? **(2)**

   (b) Copy and complete the equation below to represent the dissociation of ethanoic acid in water.

   $CH_3COOH(aq)$ ⇌ _____ (aq) + _____ (aq)

   (3)

3. Draw the structural formula of the ester ethylethanoate. **(2)**

4. (a) (i) Name the two organic substances required to make the ester ethylethanoate. **(2)**
   (ii) Name the catalyst required to speed up the reaction between them. **(2)**

   (b) The reaction to prepare ethylethanoate is carried out by heating the reactants under reflux. Draw the apparatus used to do this in the laboratory. **(4)**

Section Four

4 Physical chemistry

A Acids, alkalis and salts

You will be expected to:

★ describe the use of the some indicators to distinguish between acidic and alkaline solutions
★ explain how the pH scale can be used to classify solutions
★ describe the use of universal indicator
★ define acids as sources of hydrogen ions and alkalis as sources of hydroxide ions
★ predict the properties of some reactions of dilute hydrochloric, nitric and sulfuric acids
★ outline the general rules for predicting the solubility of salts in water
★ describe how to prepare soluble salts from acids
★ describe how to prepare insoluble salts using precipitation reactions
★ describe how to perform acid–alkali titrations.

Measuring pH

Water is neutral and has a pH equal to 7. When a substance dissolves in water the solution formed may be:

* **acidic** – pH less than 7
* **neutral** – pH equal to 7
* **alkaline** – pH greater than 7.

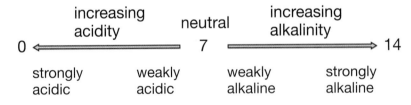

Fig 4a.01: The pH scale

Indicators

Indicators can be used to show whether a solution is acidic, alkaline or neutral by the way their colour changes, for example:

Fig 4a.02: Indicators with acidic and alkaline solutions

Universal indicator produces a large range of colours over a large range of pH, making it useful for identifying pH more accurately than the indicators in Fig. 4a.02.

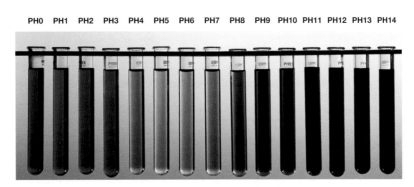

Fig. 4a.03: Range of colours of universal indicator at different pHs

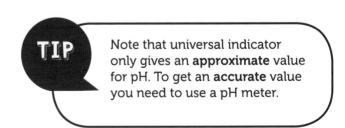

TIP

Note that universal indicator only gives an **approximate** value for pH. To get an **accurate** value you need to use a pH meter.

Acids

All acids dissolve in water to form hydrogen ions, H^+ (which are also *protons*). The other ion formed depends on the acid.

Acid	Ions formed in water
hydrochloric acid (HCl)	hydrogen ions and chloride ions: $H^+(aq)$ and $Cl^-(aq)$
nitric acid (HNO_3)	hydrogen ions and nitrate ions: $H^+(aq)$ and $NO_3^-(aq)$
sulfuric acid (H_2SO_4)	hydrogen ions and sulfate ions: $H^+(aq)$ and $SO_4^{2-}(aq)$

TIP Note that the ions formed in water are **hydrated** (i.e. have water molecules clustered around them) and this is represented by the '(aq)' written after the ion, for example $H^+(aq)$ and $Cl^-(aq)$.

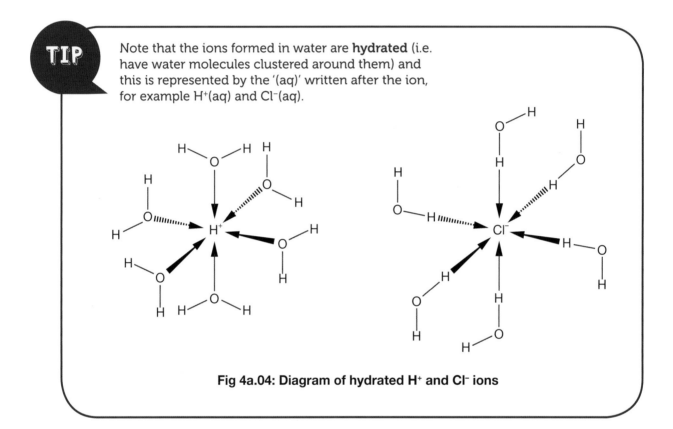

Fig 4a.04: Diagram of hydrated H^+ and Cl^- ions

Alkalis

Alkalis are sources of hydroxide ions, $OH^-(aq)$. For example, the hydroxides of the Group 1 metals (also called the alkali metals), such as sodium hydroxide (NaOH) and potassium hydroxide (KOH), dissolve in water to produce hydroxide ions, $OH^-(aq)$. Hydroxide ions make the resulting solutions alkaline.

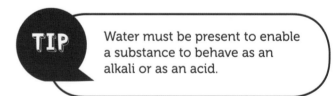

TIP Water must be present to enable a substance to behave as an alkali or as an acid.

Neutralisation

A neutral solution has a pH of 7 and can be formed by mixing an acid with an alkali in the correct proportions.

During the reaction, known as **neutralisation**, hydrogen ions from the acid are accepted by the hydroxide ions from the alkali to form water:

$$H^+(aq) + OH^-(aq) \rightarrow H_2O(l)$$

CAM

> Hydrogen ions are also known as protons. Acids are proton donors, and bases are proton acceptors. In the reaction $NaOH + HCl \rightarrow NaCl + H_2O$ the acid (HCl) donates a proton to the base (NaOH) as shown in the equation above.

Strong and weak acids and alkalis

Not all acids and alkalis dissociate in water to the same extent.

- Those that dissociate completely in water are described as being *strong*.
- Those that dissociate incompletely (about 1%) are described as *weak*.
- Strong acids include hydrochloric, sulfuric and nitric acids.
- Weak acids include organic acids, such as ethanoic acid.
- Strong alkalis include aqueous solutions of the hydroxides of sodium and potassium.
- Weak alkalis include an aqueous solution ammonia, known as ammonium hydroxide.

TIP

> For a weak acid, such as ethanoic acid (CH_3COOH), an equilibrium symbol is used to indicate that the dissolved acid still exists mainly in the form of ethanoic acid molecules:
>
> $$CH_3COOH(aq) \rightleftharpoons H^+(aq) + CH_3COO^-(aq) \quad \text{[partially dissociated]}$$
>
> This implies that aqueous solutions of a strong and a weak acid of equal concentration would have very different concentrations of hydrogen ions, and hence a different pH.
>
> The dissociation of a strong acid is shown with an arrow, not the equilibrium symbol:
>
> $$HCl(aq) \rightarrow H^+(aq) + Cl^-(aq) \quad \text{[100\% dissociated]}$$

For solutions of weak and strong acids of the same concentration:

- strong acids have a *lower* pH (greater hydrogen ion concentration) than a weak acid
- strong acids react more rapidly than weak acids with reactive metals such as zinc because the hydrogen ion concentration is greater, increasing the rate of reaction.

Reactions of acids

Reactions with metals

(See summary table on page 134)

Hydrochloric acid reacts with reactive metals (e.g. magnesium, aluminium, zinc and iron) to form hydrogen gas and metal chlorides, e.g.:

$$Zn \quad + \quad 2HCl \quad \rightarrow \quad ZnCl_2 \quad + \quad H_2$$

zinc metal + hydrochloric acid → zinc chloride + hydrogen

Sulfuric acid reacts with reactive metals (e.g. magnesium, aluminium, zinc and iron) to form hydrogen gas and metal sulfates, e.g.:

$$Fe \quad + \quad H_2SO_4 \quad \rightarrow \quad FeSO_4 \quad + \quad H_2$$

iron metal + sulfuric acid → iron sulfate + hydrogen

Reactions with metal oxides

Hydrochloric acid reacts with metal oxides to form metal chlorides and water, e.g.:

$$ZnO \quad + \quad 2HCl \quad \rightarrow \quad ZnCl_2 \quad + \quad H_2O$$

zinc oxide + hydrochloric acid → zinc chloride + water

Sulfuric acid reacts with metal oxides to form metal sulfates and water, e.g.:

$$MgO \quad + \quad H_2SO_4 \quad \rightarrow \quad MgSO_4 \quad + \quad H_2O$$

magnesium oxide + hydrochloric acid → magnesium chloride + water

Nitric acid reacts with metal oxides to form metal nitrates and water, e.g.:

$$CuO \quad + \quad 2HNO_3 \quad \rightarrow \quad Cu(NO_3)_2 \quad + \quad H_2O$$

copper oxide + nitric acid → copper nitrate + water

Reactions with metal carbonates

Hydrochloric acid reacts with metal carbonates to form metal chlorides, water and carbon dioxide, e.g.:

$$CaCO_3 \quad + \quad 2HCl \quad \rightarrow \quad CaCl_2 \quad + \quad CO_2 \quad + H_2O$$

calcium carbonate + hydrochloric acid → calcium chloride + carbon dioxide + water

Sulfuric acid reacts with metal carbonates to form metal sulfates, water and carbon dioxide, e.g.:

$$Na_2CO_3 \quad + \quad H_2SO_4 \quad \rightarrow \quad Na_2SO_4 \quad + \quad CO_2 \quad + H_2O$$

sodium carbonate + sulfuric acid → sodium sulfate + carbon dioxide + water

Nitric acid reacts with metal carbonates to form metal nitrates, water and carbon dioxide, e.g.:

$$CaCO_3 \quad + \quad 2HNO_3 \quad \rightarrow \quad Ca(NO_3)_2 \quad + \quad CO_2 \quad + H_2O$$

calcium carbonate + nitric acid → calcium nitrate + carbon dioxide + water

Chemistry A* Study Guide

Controlling the acidity of soils

Calcium carbonate, and to a lesser extent calcium hydroxide, are used to control the acidity of soils. In general, plants grow best at about pH 7. Many studies show that crop yields are affected by soil pH. The graph in Fig. 4a.05 shows the effect of pH on corn yields.

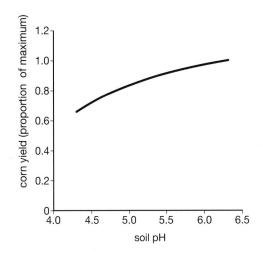

Fig. 4a.05: The effect of pH on corn yields

The following tend to lower the pH of soil:

* Fertiliser use. The ammonium ions present in fertilisers dissociate in water, releasing hydrogen ions:
$$NH_4^+(aq) \rightarrow NH_3(g) + H^+(aq)$$

* Weathering of rocks. Aluminium ions are released and react with soil water in a series of reactions, the first of which is represented by the equation:
$$Al^{3+}(aq) + H_2O(l) \rightarrow Al(OH)^{2+}(aq) + H^+(aq)$$

* Acidic rain. Rainwater is acidic owing to the presence of dissolved weak acids such as carbonic acid, H_2CO_3, and sulfurous acid, H_2SO_3

The acidity of the soil itself is not the main problem, but rather the effect this has on the availability of plant nutrients. The plot in Fig. 4a.06 shows how pH affects the availability of some essential plant nutrients.

The width of the band for each nutrient shows its availability at a given pH: nitrogen reaches its maximum availability at pH 6, while magnesium (an important constituent of chlorophyll, vital for photosynthesis) requires a pH of 7 if it is to be fully available. The elements nitrogen, phosphorus and potassium are essential for healthy plant growth and are contained in fertilisers applied to the soil. Some nitrogen reaches the soil naturally in rainfall, but the amounts are too small to allow the high cropping levels demanded by modern intensive farming.

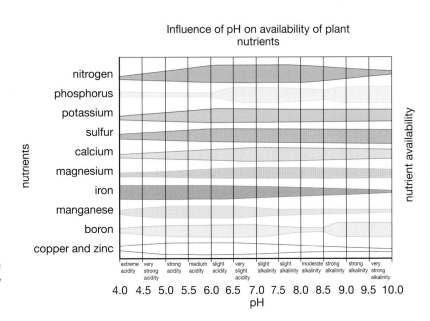

Fig. 4a.06: The influence of pH on availability of plant nutrients

Summary of metal reactions with acids

The metal ion in the salt depends on the metal, metal oxide or metal carbonate used:

Acid used	Salt formed	Example
hydrochloric	chloride	zinc + hydrochloric acid → zinc chloride
sulfuric	sulfate	copper oxide + sulfuric acid → copper sulfate
nitric	nitrate	sodium carbonate + nitric acid → sodium nitrate

Preparation of salts

A **salt** is the product of a reaction in which the hydrogen ions in an acid are replaced by a metal ion (or the ammonium ion) from a base. For example:

Parent acid	Replace H^+ with	Resulting salt
hydrochloric acid, HCl	potassium ions, K^+	potassium chloride, KCl
sulfuric acid, H_2SO_4	copper(II) ions, Cu^{2+}	copper(II) sulfate, $CuSO_4$
nitric acid, HNO_3	zinc ions, Zn^{2+}	zinc nitrate, $Zn(NO_3)_2$
hydrochloric acid, HCl	ammonium ions, NH_4^+	ammonium chloride, NH_4Cl

Solubility rules for salts

Soluble salts are:

*   all nitrates
*   all common sodium, potassium and ammonium salts
*   common chlorides *except silver chloride*
*   common sulfates *except those of calcium and barium*.

Insoluble salts are:

*   common carbonates and hydroxides *except those of sodium, potassium and ammonium*.

You will be expected to use these solubility rules to decide how best to prepare a particular salt using one of the methods described below.

Preparing soluble salts from an acid and a base

A **base** is a substance that does not dissolve in water, but reacts with an acid to form a salt and water.

The method for producing a salt depends on the fact that once all of the acid present has reacted the base will not dissolve and can be simply removed by filtration.

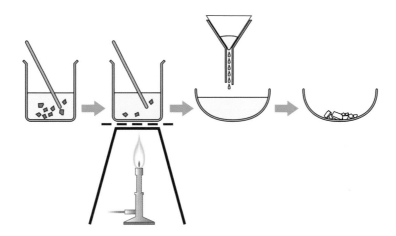

Fig 4a.07: Preparation of a salt from an acid and a base

- Add the solid base (e.g. copper(II) oxide) in small portions to about 100 cm^3 of a dilute acid (i.e. nitric acid if a nitrate is to be prepared) in a beaker.
- Warm the mixture, stir and continue to add the base until a small quantity of undissolved base remains. (This ensures that all of the *acid* has reacted.)
- Filter the hot solution: the residue on the filter paper is unreacted base, the filtrate is an aqueous solution of the salt.
- Transfer the filtrate to an evaporating basin and gently evaporate the solution (best over a water bath to avoid 'spitting' or over-heating) until the volume of the solution has been reduced to about one-third of its initial volume.
- Set the evaporating basin aside, loosely covered with a piece of filter paper to keep out dust, until crystals have formed (overnight is usually sufficient time).
- Pour off the liquid from the crystals and pat them dry between filter papers.
- Finally dry the crystals in an oven.

Note that this procedure can be used to prepare salts from an *acid and a metal*.

Preparing insoluble salts

Insoluble salts can be conveniently prepared by mixing two aqueous solutions, one containing the cation in the salt and the other containing the anion. The desired salt forms as a **precipitate** which is filtered off, washed with water and dried in an oven.

- The method is limited by the range of insoluble salts, but the following can be made in this way:
- silver halides (AgX: where X = Cl, Br or I) – mix silver nitrate solution with a solution containing the appropriate halide ion
- lead halides (PbX$_2$: where X = Cl or I) – mix lead nitrate solution with a solution containing the appropriate halide ion
- barium sulfate BaSO$_4$ – mix barium chloride solution with a solution containing the sulfate ion.

Preparing soluble salts from an acid and an alkali by titration

The method for preparing soluble salts from bases above will not work with an alkali in place of the base because the alkali would continue to dissolve in the water long after all of the acid had been neutralised. This would mean that the desired salt would be contaminated with the alkali and could not be obtained as pure crystals.

To prepare salts from acids and alkalis a technique known as **titration** must be used.

Aqueous solutions of the acid and alkali must be mixed in exactly the right quantities to ensure that the acid has been neutralised but that no excess alkali is added. An *indicator* is used to identify the point at which just enough acid and alkali have been mixed, that is when the solution is *neutral*.

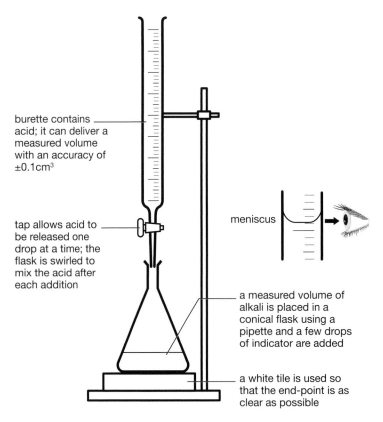

burette contains acid; it can deliver a measured volume with an accuracy of ±0.1cm³

tap allows acid to be released one drop at a time; the flask is swirled to mix the acid after each addition

meniscus

a measured volume of alkali is placed in a conical flask using a pipette and a few drops of indicator are added

a white tile is used so that the end-point is as clear as possible

Fig. 4a.08: Titration

Standard procedure for titration:

- Measure 25 cm³ of alkali into the conical flask using a pipette and add a few drops of indicator.
- Fill the burette with the acid and note the initial burette reading.
- Slowly run the acid into the alkali, swirling constantly.
- Stop adding acid when the indicator *just* changes colour.
- Note the final burette reading and calculate the volume of acid added.
- Repeat the procedure until at least two values of the volume of acid added agree to within 0.1 cm³. Average the volumes added.

To prepare the soluble salt:

- Repeat the whole experiment, omitting the indicator: i.e. simply add the volume of acid calculated to 25 cm^3 of alkali. This gives a neutral solution of the salt uncontaminated by indicator.
- Place the resulting solution in an evaporating basin and reduce its volume to one-third by careful evaporation over a boiling water bath.
- Set the hot solution aside to cool until crystals form.
- Decant (pour off) the solution from the crystals and dry them between filter papers and then dry in an oven.

You should now be able to:

★ describe the colours of the indicators litmus, phenolphthalein and methyl orange in acidic and in alkaline solutions (see page 129)

★ give the range of pH for strongly acidic, weakly acidic, neutral, weakly alkaline and strongly alkaline solutions (see page 133)

★ explain how to measure the approximate pH of a solution using universal indicator (see page 129)

★ define the terms *acid* and *alkali* in terms of the ions they produce in aqueous solution (see page 130)

★ name the products formed when the following substances react: hydrochloric acid and zinc; sulfuric acid and copper(II) oxide; nitric acid and sodium carbonate (see page 132)

★ predict the solubility of the following salts: sodium sulfate; calcium carbonate; silver chloride; barium sulfate (see page 134)

★ describe how to prepare potassium nitrate from nitric acid (see page 135)

★ describe how to prepare silver bromide from silver nitrate solution (see page 135)

★ describe how to carry out an acid–alkali titration accurately (see page 136).

Practice questions

1. (a) Explain how these pieces of equipment are used during a titration.

    (i) burette   (ii) conical flask   (iii) pipette   **(3)**

    (b) What is the purpose of the indicator in a titration?   **(1)**

2. (a) Write equations to show how the following acids dissociate in water.

    (i) hydrochloric acid, HCl   (ii) ethanoic acid, CH_3COOH **(3)**

    (b) Explain the difference between a weak and a strong acid.   **(2)**

    (c) Which of the two acids above is strong?   **(1)**

3. Copy and complete these statements.

    Acidic pH is _____ 7

    Alkaline pH is _____ 7   **(2)**

4. (a) For each of these indicators, state what colour they are in acidic solutions, and in alkaline solutions.

    (i) litmus   (ii) phenolphthalein   (iii) methyl orange   **(3)**

    (b) What colour would litmus turn if you added it to:

    (i) lemon juice   (ii) ammonia?   **(2)**

5. Copy and complete the equation below, which represents a neutralisation reaction. Include state symbols.

    _____ (_____) + _____ (_____) $\rightarrow$ H_2O(_____)   **(3)**

6. Copy and complete the table below. (6)

Acid	Alkali	Salt formed
	sodium hydroxide	sodium nitrate
sulfuric acid		sodium sulfate
hydrochloric acid	sodium hydroxide	
nitric acid		potassium nitrate
	potassium hydroxide	potassium sulfate
	potassium hydroxide	potassium chloride

7. Write the corresponding *chemical equation* for each of the following word equations.

    (i) magnesium + sulfuric acid $\rightarrow$ magnesium sulfate + water
    (ii) copper oxide + nitric acid $\rightarrow$ copper nitrate + water
    (iii) zinc carbonate + hydrochloric acid $\rightarrow$ zinc chloride + water + carbon dioxide   **(10)**

8. Copy the table on the following page and tick the boxes to show which salts are soluble and which are insoluble.   **(6)**

Salt	Soluble?	Insoluble?
sodium bromide		
ammonium sulfate		
silver chloride		
potassium carbonate		
barium carbonate		
lead nitrate		

9. (a) Write a *balanced symbol equation*, with state symbols, for the reaction between copper(II) oxide, CuO, and sulfuric acid, H_2SO_4. (3)

   (b) Write the *ionic equation*, with state symbols, for the reaction between CuO and H_2SO_4. (3)

10. This question is concerned with different methods of preparing salts.

   Method A: Reaction of an acid and alkali (metal hydroxide or ammonia solution) by **titration** using a burette and an indicator.

   Method B: Add an *excess* of a *solid base* or *metal* or *carbonate* to a dilute acid and remove the excess by filtration.

   Method C: Mix two solutions and obtain an insoluble salt by precipitation. Filter off the salt, wash and dry the residue.

   Choose one of these methods for making salts in the following questions. In each case give a word equation for the reaction.

   (a) To make copper sulfate using solid copper oxide. (2)
   (b) To make potassium chloride using potassium hydroxide. (2)
   (c) To make the insoluble salt silver bromide. (2)
   (d) To make sodium bromide using sodium hydroxide. (2)
   (e) To make zinc chloride using zinc oxide. (2)
   (f) To make the insoluble salt barium carbonate. (2)
   (g) To make the insoluble salt lead iodide. (2)
   (h) To make calcium nitrate using calcium carbonate. (2)
   (i) To make zinc chloride using zinc. (2)

11. Describe how you would prepare copper sulfate crystals from copper oxide in the laboratory. (8)

12. Copy and complete the table below. (8)

Solutions mixed	Precipitate formed? (Yes/No)	Name of precipitate
sodium chloride and silver nitrate		
sulfuric acid and barium nitrate		
sodium carbonate and calcium nitrate		
potassium hydroxide and calcium chloride		

13. Name two groups of salts, all of which are soluble in water. (2)

B Energetics

You will be expected to:

★ explain the meaning of the terms *exothermic* and *exothermic*
★ describe simple laboratory experiments in which heat changes can be calculated from measured temperature changes
★ calculate molar enthalpy changes
★ describe the use of ΔH to represent enthalpy changes in reactions
★ represent exothermic and endothermic reactions on a simple energy level diagram
★ explain that bond breaking is endothermic and that bond forming is exothermic
★ use average bond energies to calculate enthalpy changes in reactions.

Exothermic reactions

Exothermic changes *give out* heat to the surroundings.

• They require energy to be supplied to start them (the **activation energy**) (see Section 4c).
• They keep on going until one of the reagents is consumed.

> **TIP** Learn the following examples of exothermic reactions for the examination:
>
> • combustion of fuels, e.g. methane
> • respiration (the conversion of food molecules to carbon dioxide and water, releasing energy)
> • hydration of anhydrous copper(II) sulfate
> • hydration of concentrated sulfuric acid
> • neutralisation of an acid by an alkali.

The heat energy becomes available because during the reaction more energy is released when the new bonds form in the products than is required to break the bonds in the reactants. The difference is the heat energy given out.

For example, when methane burns:

$$CH_4(g) + 2O_2(g) \rightarrow CO_2(g) + 2H_2O(l)$$

Fig 4b.01: Bond breaking and formation in methane combustion

Weak C–H and O=O bonds break to be replaced in the products by strong C=O bonds and O–H bonds.

During the process the reactants (CH_4 and O_2) **lose** energy. This can be indicated by making an addition to the equation as shown below:

$$CH_4(g) + 2O_2(g) \rightarrow CO_2(g) + 2H_2O(l) \qquad \textbf{ΔH} = \underline{\textbf{–890 kJ mol}^{-1}}$$

- The symbol ΔH means *heat change*.
- The figure of 890 kJ mol⁻¹ indicates how large the heat change is for every mole of methane that is burnt.
- The minus sign is added to show that heat has been *given out*, in other words that the reaction mixture has *lost* energy to the surroundings (e.g. to a saucepan on a domestic cooker).
- The sign refers to the heat change *of the reaction mixture* as this is generally of more interest to chemists than where the heat eventually goes.

CAM

Energy level diagram for exothermic reactions

An energy level diagram for an exothermic reaction shows the initial input of activation energy, with an output of energy from the reaction that is greater. Overall, energy is given out from the reaction.

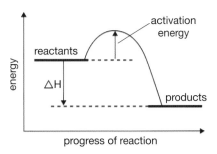

Fig 4b.02: Exothermic energy level diagram

Endothermic reactions

Endothermic changes *take in* heat from the surroundings.

- They require energy to be supplied to start them (the activation energy).
- They stop if the energy supply is withdrawn.

TIP Learn the following examples of endothermic reactions for the examination:

- photosynthesis, in which water and carbon dioxide react in the presence of a catalyst (chlorophyll) and light energy to produce sugars and oxygen.
- dissolving of some salts in water, e.g. NH_4Cl and KNO_3
- thermal decomposition of calcium and copper carbonates.

We must supply the additional energy to make the useful chemicals in this kind of reaction from another source. ΔH for such reactions is *positive*, e.g. +35 kJ mol⁻¹.

Energy level diagram for endothermic reactions

An energy level diagram for an endothermic reaction shows the initial input of activation energy, with an output of energy from the reaction that is less than this. Overall, energy is taken in during the reaction.

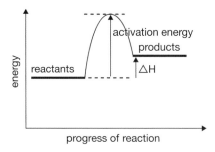

Fig. 4b.03: Endothermic energy level diagram

Measuring heat changes in the laboratory

We cannot measure heat changes directly but can work them out from the temperature change which occurs when a known mass of a substance (usually water) is heated by the reaction. We use the equation:

heat change of reaction = mass of water × specific heat capacity of water × temperature change

By convention, we show heat given out (in an exothermic reaction) as a negative (-) quantity and heat taken in (in an endothermic reaction) as a positive (+) quantity.

Specific heat capacity

The **specific heat capacity** of a substance is the amount of heat energy that must be supplied to raise the temperature of 1 gram by 1 °C. For water, this is 4.2 J °C^{-1} g^{-1}.

*   To raise the temperature of 1 g of water by 10 °C requires 4.2 × 10 = 42 joules.
*   To raise the temperature of 10 g of water by 10 °C requires 4.2 × 10 × 10 = 420 joules.
*   To raise the temperature of 5 g of water by 2 °C requires 4.2 × 5 × 2 = 42 joules.

TIP

You will be expected to know how to measure the heat change when:

*   a fuel is burned (the heat of combustion)
*   a metal ion is displaced from its salt by another metal (the heat of displacement)
*   a salt dissolves in a solution (the heat of solution)
*   an acid is neutralised by an alkali (the heat of neutralisation).

The final answer must include a sign to indicate whether the reaction is endothermic (+) or exothermic (-)

Measuring heat of combustion

A standard procedure is:

- fill a burner with ethanol and record its mass
- place a known mass of water in a beaker and clamp it over the burner
- record the temperature of the water
- light the burner and let it burn until the temperature of the water has risen by 10–20 °C, noting the final temperature
- blow out the burner, let it cool, record its mass and calculate the mass of ethanol burned.

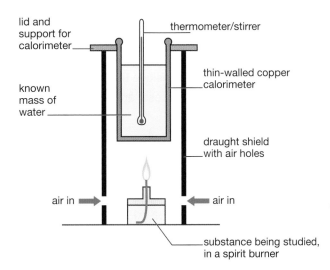

Fig 4b.04: Equipment for measuring heat of combustion

Worked example

Some typical results might be:

- initial mass of burner   56.90 g
- final mass of burner   56.70 g
- mass of ethanol burned   0.20 g
- final temperature of water   36.2 °C
- initial temperature of water   22.0 °C
- temperature rise   14.2 °C
- mass of water in beaker   100 g

So:

mass of ethanol burned = 0.20 g [M_r ethanol 46]

moles of ethanol burned = $0.20 \div 46 = 4.35 \times 10^{-3}$ moles

heat evolved by 4.35×10^{-3} moles of ethanol:

= $100 \times 4.2 \times 14.2$ joules

= 5964 joules

heat of combustion of 1 mole of ethanol:

= $5964 \div (4.35 \times 10^{-3})$

= 1371×10^3 joules mol^{-1} = **-1371 kJ mol^{-1}**

The main source of error in this experiment is that not all of the heat given off on combustion of the ethanol goes to heat the water. Some 'escapes' around the beaker, some heats the air, etc. A copper can is often used instead of a beaker because it conducts heat better than glass and can be made thinner and hence can have a lower specific heat capacity.

Measuring heat energy change in displacement reactions

CAM

The following method for the reaction between a copper salt and zinc metal can be used to measure the heat energy change in all displacement reactions. The reaction takes place in a polystyrene beaker to insulate it from the surroundings.

To measure the heat change that occurs when known quantities of aqueous copper(II) ions and zinc metal react according to the equation:

$$Cu^{2+}(aq) + Zn(s) = Cu(s) + Zn^{2+}(aq)$$

- place 25 cm³ of 0.2M copper(II) sulfate solution in a polystyrene beaker
- record the initial temperature ($T_{initial}$) using a sensitive thermometer
- weigh out accurately 0.5 g of zinc powder (an excess) and add it all quickly to the copper(II) sulfate solution stirring as you do so
- record the highest final temperature reached (T_{final}).

Very little error is introduced if it is assumed that the specific heat capacity of the copper(II) sulfate solution is 4.2 J g⁻¹ °C⁻¹, (i.e. 4.2 J of heat raise the temperature of 1 g of it by 1 °C) and that the density of the solution is 1 g cm⁻³ (i.e. the same values as for pure water).

It is safe to assume that the heat given out is absorbed only by the solution and not the polystyrene beaker, because this has a small mass and a low specific heat capacity. So little inaccuracy is caused.

Worked example for copper salt and zinc metal

(Relative atomic mass of zinc = 65)

Mass of zinc used = 0.45

> Note that the zinc is in excess, so some unreacted metal remains at the end of the reaction, but this can be neglected in the calculations as shown below.

Moles of zinc used = 0.45 ÷ 65
$$= 6.9 \times 10^{-3}$$

Number of moles of copper(II) ions involved = $25 \times 0.2 \times 10^{-3} = 5 \times 10^{-3}$

- Initial temperature ($T_{initial}$) = 23.1 °C
- Final temperature (T_{final}) = 32.5 °C

Temperature change = $T_{final} - T_{initial}$ = 32.5 − 23.1 = 9.4 °C

Heat change = specific heat capacity × mass × temperature change

$$= 4.2 \times 25 \times 9.4 = 987 \text{ J}$$

> Note that the 'mass' referred to in this equation is the mass of the solution used, which is assumed to have the same density and specific heat capacity as water.

Chemistry A* Study Guide

The heat change which would occur when 1 mole of copper(II) ions are displaced = heat change ÷ moles of copper(II) ions used:

$$= 987 \div (5 \times 10^{-3}) = 197 \times 10^3 \text{ J mol}^{-1} = 197 \text{ kJ mol}^{-1}$$

The number of moles of unreacted zinc = $(6.9 - 5) \times 10^{-3} = 1.9 \times 10^{-3}$ moles.

The mass of 1.9×10^{-3} moles of Zn = $65 \times 1.9 \times 10^{-3}$ g = 0.12 g.

The specific heat capacity of Zn = 0.39 J/g, so the heat absorbed in raising its temperature by 9.4 °C

$$= 0.12 \times 0.39 \times 9.4 = 0.44 \text{ J}$$

This is negligible when compared with the enthalpy change for the displacement reaction of 987 J.

Measuring heat energy change in dissolving

The experimental procedure, using a polystyrene cup as a calorimeter, is the same as that for enthalpy of displacement (above). The worked example below determines the enthalpy of dissolving (solution) of potassium nitrate, KNO_3.

Worked example

(Relative atomic masses: K = 39; N = 14; O = 16)

- molar mass of KNO_3 =   101 g mol^{-1}
- mass of KNO_3 used   =   25.25 g
- moles KNO_3 used   =   25.25 ÷ 101 = 0.25 mol

When 25.25 g of KNO_3 was dissolved in 100 cm³ of water the temperature decreased by 20.5 °C.

Heat change = specific heat capacity × mass × temperature change (using values for s h c and mass for water as before).

$$= 4.2 \times 100 \times (20.5) = 8610 \text{ J}$$

The heat change which would occur when 1 mole of potassium nitrate dissolves = heat change ÷ moles of potassium nitrate used.

$$= 8610 \div 0.25 = 34\,440 \text{ J mol}^{-1} = \textbf{34.4 J mol}^{-1}$$

Measuring heat of neutralisation

A standard procedure is:

- place 25.0 cm³ of acid of known concentration (e.g. 2.0 mol dm⁻³) in a polystyrene beaker and record its temperature with a thermometer
- place 25.0 cm³ of alkali of known concentration (e.g. 2.0 mol dm⁻³) in a glass beaker and record its temperature with a thermometer (if the two temperatures differ, take the average)
- quickly pour the alkali into the acid in the polystyrene beaker, stir to mix and record the highest temperature reached.

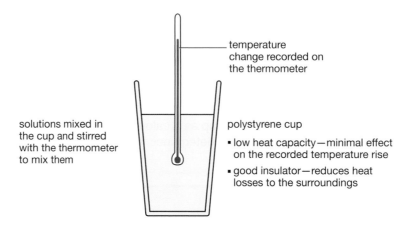

temperature change recorded on the thermometer

solutions mixed in the cup and stirred with the thermometer to mix them

polystyrene cup
- low heat capacity—minimal effect on the recorded temperature rise
- good insulator—reduces heat losses to the surroundings

Fig. 4b.05: Equipment for measuring heat of neutralisation

Worked example

Some typical results might be:

- volume of acid used   25.0 cm³
- volume of alkali used   25.0 cm³
- concentration of acid and alkali   2.0 mol dm⁻³
- initial temperature   21.0 °C
- final temperature   34.8 °C
- temperature rise   13.8 °C

As before, we can assume that the resulting solution has the same specific heat capacity as water (4.2 J g⁻¹ °C⁻¹) and that all solutions have the same density as water i.e. 1.0 g cm⁻³.

So:

moles of acid neutralised = volume × conc. in mol dm⁻³ × 10⁻³

$$= 25.0 \times 2.0 \times 10^{-3} = 0.05 \text{ moles}$$

heat given out when this number of moles of acid are neutralised:

$$= 4.2 \times 50.0 \times 13.8 \text{ J}$$
$$= 2898 \text{ J}$$

Note we use *50* here because 25 cm³ *each* of acid and alkali were used, so the total volume is 50 cm³.

Heat given out when 1 mole of acid is neutralised:

$$= 2898 \div 0.05 \text{ J}$$

$$= \underline{-57\,960 \text{ J mol}^{-1}} \text{ or } \underline{-57.96 \text{ kJ mol}^{-1}}$$

Enthalpy changes from bond energies

All chemical reactions can be thought of as taking place in two separate stages, which involve breaking of bonds in the reagents (*endothermic*), followed by the formation of new bonds in the products (*exothermic*).

If the energy needed to *break* the bonds in the reagents is less than that given out when bonds *form* in the products, the overall reaction is *exothermic*, and vice versa for endothermic reactions.

Consider the reaction between hydrogen and chlorine to give hydrogen chloride:

$$H_2(g) + Cl_2(g) \rightarrow 2HCl(g)$$

The energy required to break the bonds involved are:

H–H 436 kJ mol^{-1} Cl–Cl 243 kJ mol^{-1} H–Cl 432 kJ mol^{-1}

Note that the values refer to the energy required to *break one mole* of the bond specified.

It should be obvious that the energy *given out* when the two atoms forming a bond combine together is equal to the energy required to break the bond, but with an opposite sign, e.g.

436 kJ taken in	436 kJ given out
bond breaking: $H_2(g) \rightarrow 2H(g)$	bond forming: $2H(g) \rightarrow H_2(g)$
+436 kJ	–436 kJ

Calculating heat changes from bond energies

Worked examples

Example 1

Calculate the heat change when the reaction represented by the following equation takes place:

$$H_2 + Cl_2 \rightarrow 2HCl$$

Answer

Step 1: Add together the energy required to *break* all the bonds in the reagents:

H–H	+436 kJ
Cl–Cl	+243 kJ

Total energy taken in =   +679 kJ

Step 2: Add together the energy given out when all the bonds in the products *form*:

2H–Cl	2 × 432 = –864 kJ

Total energy given out =   –864 kJ

Step 3: Add the energies from steps 1 and 2 together *with their signs*:

+679 + (–864) = -185 kJ (exothermic)

The resulting value is the heat change accompanying the reaction and the sign indicates whether the reaction is exothermic (–) or endothermic (+).

The result shows that, when one mole of hydrogen reacts with one mole of chlorine to form two moles of hydrogen chloride, 185 kJ of heat are given out.

Example 2

Calculate the heat change when the reaction represented by the following equation takes place:

$$N_2(g) + 3H_2(g) \rightarrow 2NH_3(g)$$

Answer

Bond strengths in kJ mol^{-1}: N≡N 945; H–H 436; N–H 391

Step 1: The energy required to break all the bonds in the reagents is:

N≡N		+945 kJ
3H–H	3 × 436 =	+1308 kJ

Total energy taken in = +2253 kJ

Step 2: The energy given out when all the bonds in the products form is:

6N–H	6 × 391 =	–2346 kJ

Total energy given out + –2346 kJ

Step 3: Add the energies from steps 1 and 2 together *with their signs*:

$$+2253 + (–2346) = \textbf{–93 kJ (exothermic)}$$

When two moles of ammonia are formed from nitrogen and hydrogen, 93 kJ of heat are given out.

Example 3

Calculate the heat change when the reaction represented by the following equation takes place:

$$CH_2=CH_2(g) + H_2(g) \rightarrow CH_3–CH_3(g)$$

Answer

Bond strengths in kJ mol-1: C=C 612; C–C 347; C–H 413; H–H 436

Note that in this reaction some of the bonds in the reagents survive intact in the product (the four C–H bonds in the ethene) so there is no need to consider them when calculating the heat change for the reaction.

Step 1: The energy required to break those bonds in the reagents *which do not exist* in the product is:

C=C		+612 kJ
H–H		+436 kJ

Total energy taken in = +1048 kJ

Step 2: The energy given out when all the bonds in the products form is:

C–C		–347 kJ
2C–H	2 × 413 =	–826 kJ

Total energy given out = –1173 kJ

Step 3: Add the energies from steps 1 and 2 together *with their signs*:

$$+1048 + (–1173) = \textbf{–125 kJ (exothermic)}$$

When one mole of hydrogen molecules adds across the double bond in ethene, 125 kJ of heat are given out.

You should now be able to:

★ define the terms *exothermic* and *endothermic* (see pages 140, 141)

★ explain the use of ΔH for indicating heat changes (see page 141)

★ describe how to measure the heat change which occurs during (a) a combustion reaction, (b) a displacement reaction, (c) dissolving, (d) a neutralisation reaction (see pages 143, 144, 145, 146)

★ explain how to convert measured heat changes into a molar enthalpy change (see page 143)

★ draw an enthalpy level diagram for the combustion of methane (see page 141)

★ explain how to calculate the enthalpy change for a reaction given the average bond enthalpies of the bonds broken and formed in the reaction (see page 147).

Practice questions

1. Copy and complete the following sentence.

   In an exothermic reaction _____ bonds in the reactants are broken and _____ bonds are formed in the products. **(2)**

2. (a) Sketch an energy level diagram for an *endothermic reaction*.
      Mark on your diagram (i) the activation energy and (ii) the heat change during the reaction. **(5)**

   (b) Name an important endothermic reaction. **(1)**

3. This question concerns the reaction represented by the word equation below:

   hydrated copper sulfate + heat energy ⇌ anhydrous copper sulfate + water
   (The symbol ⇌ indicates that the reaction is reversible.)

   (a) (i) If the energy change when the reaction takes place from left to right is +Q, what is the energy change for the right to left reaction? **(1)**
      (ii) Explain your answer to (a) (i). **(2)**

   (b) (i) What *two* observations would you expect to make when the reaction takes place in the forward direction, i.e. from left to right? **(2)**

   (c) In which direction L → R or R → L is the reaction *endothermic*? Explain your answer. **(2)**

4. The diagram shows a reaction profile for a reaction.

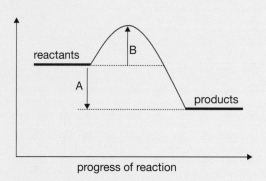

   (a) What label should appear on the vertical axis of the diagram? **(1)**

   (b) What is represented by the following on the diagram:
      (i) arrow A   (ii) arrow B? **(2)**

5. Methane burns according to the equation $CH_4(g) + 2O_2(g) \rightarrow CO_2(g) + 2H_2O(g)$.

   The total energy needed to break the bonds in methane and oxygen is 2640 kilojoules, while the energy given out when the bonds in carbon dioxide and water form is 3338 kilojoules.

   (a) Calculate the enthalpy change for the reaction. **(2)**

   (b) Is the reaction exothermic or endothermic? Explain your answer. **(2)**

6. Use the average bond energies in the table to calculate the enthalpy changes for the reactions below.

Bond broken	Bond energy (kJ mol^{-1})
C–C	348
C–H	412
C=C	612
O=O	496
H–H	436
O–H	463
C=O	743
C–O	360

(a) $2C_2H_6(g) + 7O_2(g) \rightarrow 4CO_2(g) + 6H_2O(g)$ **(4)**

(b) $CH_3CH_2OH(g) + 3O_2(g) \rightarrow 2CO_2(g) + 3H_2O(g)$ **(4)**

(c) $CH_4(g) + 2H_2O(g) \rightarrow CO_2(g) + 4H_2(g)$ **(4)**

(d) $CH_3.CH=CH_2(g) + H_2(g) \rightarrow CH_3CH_2CH_3(g)$ **(4)**

(e) $CH_3(CH_2)_3CH_3(g) \rightarrow CH_2=CH_2(g) + CH_3CH_2CH_3(g)$ **(4)**

7. In an experiment to determine the enthalpy of combustion of methanol (CH_3OH) the following results were obtained. [A_r: C 12; H 1; O 16]

- mass of methanol burnt = 0.21 g
- mass of water heated = 100 g
- initial temperature of the water = 23.0 °C
- final temperature of the water = 34.1 °C
- specific heat capacity of water = 4.2 J g^{-1} °C^{-1}

(a) Write a balanced equation for the complete combustion of methanol. **(3)**

(b) Calculate the enthalpy of combustion of one mole of methanol using the data above. **(5)**

8. 200 cm^3 of 1.0 M copper(II) sulfate solution were placed in polystyrene cup. When the solution had reached a steady temperature of 18.5 °C, 7.00 g of powdered iron were added and the mixture was stirred rapidly. The highest temperature reached was 41.5 °C.

The equation for the reaction is $Cu^{2+}(aq) + Fe(s) \rightarrow Cu(s) + Fe^{2+}(aq)$

(a) (i) Calculate the number of moles of Cu^{2+} **(1)**

    (ii) Calculate the number of moles of Fe **(1)**

    (iii) Which reagent is in excess? **(1)**

(b) Calculate the molar enthalpy change for this reaction. (The specific heat capacity of water is 4.2 J g^{-1} °C^{-1}) **(3)**

(c) State two assumptions which are made in this calculation. **(2)**

9. The following results refer to an experiment to determine the enthalpy of solution of ammonium chloride.

    - mass of ammonium chloride used   5.4 g
    - mass of water used   100 g
    - initial temperature   20.55 °C
    - final temperature   17.10 °C

    (a) Calculate the molar enthalpy of solution of ammonium chloride in kJ mol^{-1}.   (4)

    (b) Is the reaction exothermic or endothermic? Explain your answer.   (2)

10. 50 cm^3 of hydrochloric acid and 50 cm^3 of aqueous ammonia solution were mixed. Both substances had a concentration of 0.5M. During the experiment the temperature rose by 0.31 °C. The specific heat capacity of water is 4.2 J g^{-1} °C^{-1}

    Calculate the molar enthalpy of neutralisation of hydrochloric acid by ammonia solution.   (3)

C Rates of reaction

Activation energy (E_a)

The activation energy is the *minimum* energy that the reactants must have in order to react. Consider a Bunsen burner – the gas does not burn unless energy in the form of heat is supplied to start the reaction.

Other forms of energy apart from heat, such as light, can also be used to supply the activation energy.

The activation energy, E_a, can be shown on a reaction profile, as in Fig. 4c.01, which uses the reaction

$H_2 + Cl_2 \rightarrow 2HCl$ as an example.

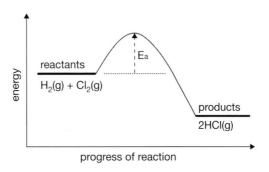

Fig. 4c.01: Activation energy graph

Rate of reaction

The rate of a reaction can be defined as:

* the rate at which the concentration of the reactants decreases with time
* or the rate at which the concentration of the products increases with time.

It can only be determined experimentally. The rate of a reaction at any time can be determined from the slope (gradient) of a plot of the change in concentration of a reactant or product at the chosen time (see Fig. 4c.02).

- The steeper the slope, the faster the rate.

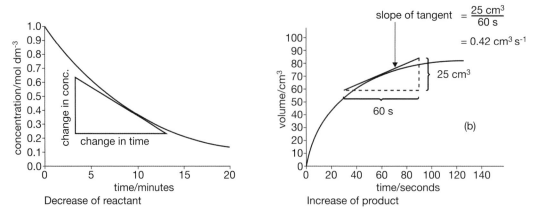

Fig. 4c.02: Calculation of rate of reaction from graphs

Example

Fig. 4c.03 shows the volume of oxygen formed as a function of time with a catalyst and in the absence of a catalyst. The reaction with a catalyst is faster and the slope of the curve with a catalyst is greater.

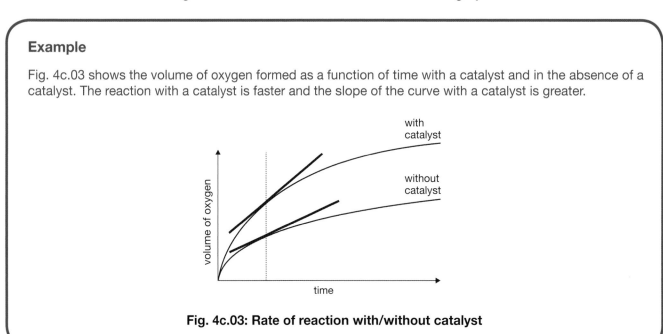

Fig. 4c.03: Rate of reaction with/without catalyst

At a given time the slope of the line for the reaction with a catalyst is greater than the slope of the line for the reaction without a catalyst showing that a catalyst increases the rate of the reaction.

Factors affecting the rate of reaction

The rate at which a chemical reaction proceeds is affected by the following:

- size of the reacting particles (surface area for reaction) in reactions involving solids
- concentration of reacting solutions
- pressure (concentration) of reacting gases
- temperature
- presence of a catalyst.

Surface area

Reactions with solids can occur only on the surface of the solid, so:

- the greater the surface area the greater the rate of reaction
- in terms of particle collision theory, increasing the surface area where reaction can take place increases the chances of collision, and so reaction, between the reactant particles.

The easiest way to increase the surface area is to grind up the solid into smaller particles.

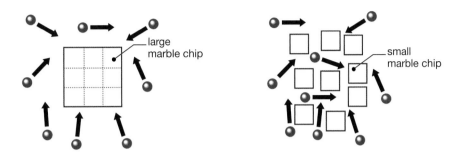

Fig. 4c.04: Effect of surface area

Example

A laboratory experiment that shows the effect of particle size uses the reaction between calcium carbonate and hydrochloric acid:

$$CaCO_3(s) + 2HCl(aq) \rightarrow CaCl_2(aq) + CO_2(g) + H_2O(l)$$

- The reaction can be carried out with powdered calcium carbonate and with lumps of calcium carbonate.
- Rate of reaction is most conveniently followed by measuring the volume of carbon dioxide given off as a function of time.
- The reaction vessel, containing solid calcium carbonate and excess hydrochloric acid, is connected to a gas syringe and the reading on the syringe is recorded as the reaction proceeds (see Fig. 4c.05).

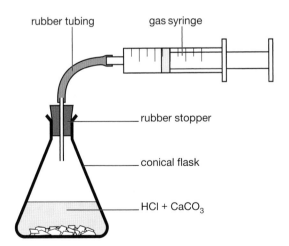

Fig. 4c.05: Calcium carbonate/acid reaction apparatus

- A graph of volume of carbon dioxide against time for each form of calcium carbonate is shown in Fig. 4c.06.

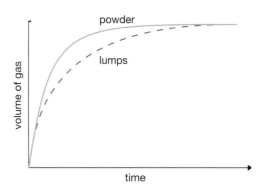

Fig.4c.06: Rate of calcium carbonate reaction

Notice that until the reaction is complete and no more gas is given off (where the graphs level off):

- the reaction slows down with time as the calcium carbonate and acid are consumed
- in a given time a greater volume of carbon dioxide is given off from the powdered calcium carbonate than from the lumps
- the *slope of the curves* represents the **rate of reaction** at any time, and this is greater for the powder than the lumps.

We can see that the mass of calcium carbonate used in the two experiments was identical, because the final volume of carbon dioxide collected is the same for both.

CAM

Reactions involving very fine powders, for example flour, can proceed with explosive violence, and there have been instances when whole flour mills have been destroyed as a consequence of the rapid (explosive) combustion of the flour grains in oxygen.

The ultimate small particles are atoms and molecules, and reactions between molecules of gas, such as methane and oxygen, can also proceed so rapidly as to cause explosions. As a result of improved detection and safety procedures explosions in coal mines involving methane and oxygen are thankfully rare, but can lead to great loss of life.

Concentration of solutions

- Increasing the concentration of a solution means that there are more reactant particles present in a given volume.
- This increases the chances of collision between them and so the reaction rate increases.

In Fig. 4c.07, the particles of the reactants are represented by blue and red balls. If the concentration of red balls is increased the reactants are more likely to collide, and the rate of reaction increases.

lower concentration of red balls   higher concentration of red balls

Fig. 4c.07: Concentration of particles

Example

A laboratory experiment that shows the effect of solution concentration on the rate of reaction uses the reaction between sodium thiosulfate and hydrochloric acid:

$$Na_2S_2O_3(aq) + 2HCl(aq) \rightarrow H_2O(l) + SO_2(aq) + S(s)$$

During this reaction solid sulfur is produced, which makes the originally transparent reaction mixture increasingly opaque. If the total volume of the reaction mixture is kept constant, the rate of reaction can be followed by measuring the time taken for a black cross placed beneath the reaction mixture to disappear.

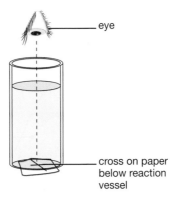

Fig. 4c.08: Thiosulfate reaction apparatus

- The shorter the time taken for the cross to disappear the faster the rate of reaction.
- So the reaction rate is *inversely proportional* to the time for the cross to disappear.

Pressure of gases

- Increasing the pressure of a gas increases the number of particles in a given volume – its concentration increases.
- In reactions involving gases, the rate of reaction is increased when the pressure of the gases is increased.

An example of this can be seen in the Haber process (see Section 5d).

Temperature

- Increasing the temperature increases the energy of the reactant particles.
- This means more of the reactant particles will have energies greater than the activation energy and so be able to react when they collide.

The reaction between sodium thiosulfate and hydrochloric acid can also be used to demonstrate the effect of temperature on rate of reaction, by carrying out the reaction described above in water baths at different at temperatures.

Presence of a catalyst

A **catalyst** is a chemical that increases the rate of a reaction but remains unchanged at the end of the reaction.

*   A catalyst lowers the activation energy for a reaction by providing an alternative reaction pathway which has a lower activation energy than the original reaction.
*   This means that at a given temperature more reactant particles will have energies in excess of the lowered activation energy.

An analogy is lowering the height of a high jump bar: at 2 m, few competitors (reactants) would have sufficient energy to clear it, but if it was lowered to 1 m almost all competitors (reactants) could do so.

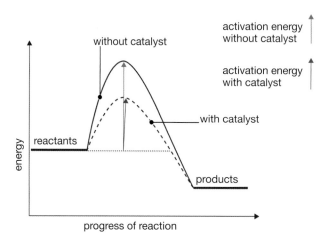

Fig 4c.09: Activation energy modified by a catalyst

Example

A laboratory experiment that shows the effect of a catalyst on rate of reaction uses the decomposition of hydrogen peroxide. Hydrogen peroxide decomposes slowly into water and oxygen:

$$2H_2O_2(aq) \rightarrow 2H_2O(l) + O_2(g)$$

The rate of decomposition is increased by a catalyst, such as manganese(IV) oxide (manganese dioxide).

The rate of reaction can be followed by measuring the volume of oxygen given off as a function of time, using the same apparatus as for the calcium carbonate/hydrochloric acid reaction.

Enzymes

Enzymes are naturally occurring catalysts, produced by living organisms, which increase the rates of biological reactions. They are damaged by extremes of temperature and pH.

Unlike other catalysts, they have an **optimum temperature** at which they work most efficiently. The optimum temperature varies from one enzyme to another, but many are **denatured** (destroyed) at temperatures above about 50 °C.

Most enzymes also have an **optimum pH** and work best over a particular pH range. So mixtures called **buffer solutions** are used to keep the pH within acceptable limits during reactions.

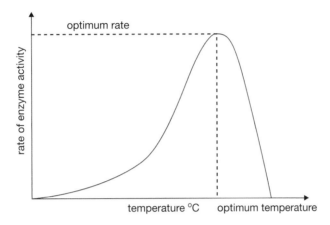

Fig. 4c.10: Enzyme activity is best at an optimum temperature

Reactions using enzymes can be carried out at lower temperatures and pressures than those using industrial catalysts. Enzyme-catalysed reactions are also very much faster. This reduces expenditure on costly equipment and also saves energy. For example, enzymes are extracted from living cells and used in washing detergents, which saves energy on heating the water for washing and the time spent washing.

Light and rate of reaction

Light is a form of energy and can cause chemical reactions to occur by providing the activation energy. It can also increase the rate of reaction if the intensity of light increases. Photography, in which light activates silver bromide crystals in photographic film that then react to give metallic silver in the presence of the developer, is an important example of the use of light energy. During the formation of the photographic image silver ions have been reduced to metallic silver.

Photosynthesis, in plants, is driven by light energy from the Sun. Carbon dioxide and water react to give sugars and oxygen in the presence of chlorophyll and light.

You should now be able to:

★ describe an experiment to investigate the effect on the rate of reaction of: (a) the surface area of a solid reactant, and (b) the presence of a catalyst (see pages 155, 158)

★ define the term activation energy and show it on an energy level diagram (see page 153)

★ explain the effects of changes in the concentration of solutions and the temperature on the rate of a reaction in terms of particle collision theory (see pages 156, 157)

★ explain how a catalyst increases the rate of a reaction (see page 158).

Practice questions

1. State and explain, in terms of particle collision theory, the effect of each of the following on the rate of a chemical reaction:

   (a) an increase in temperature **(4)**

   (b) a decrease in concentration in liquids or gases **(4)**

   (c) a reduction in particle size of solids **(4)**

   (d) the addition of a catalyst. **(4)**

2. Calcium carbonate reacts with dilute hydrochloric acid to form carbon dioxide gas. The gas escapes and the mass of the reaction vessel decreases with time.

   The rate of evolution of carbon dioxide can be followed by measuring the mass of the conical flask in which the reaction occurs at various times.

   The table below gives the results from a reaction when 40 cm^3 dilute hydrochloric acid was added to one marble chip (calcium carbonate) at 20 °C. The calcium carbonate was in excess.

Mass of flask and contents (g)	71.00	70.74	70.54	70.40	70.30	70.26	70.22	70.20	70.20
Time (mins)	0	1	2	3	4	5	6	7	8

   (a) (i) Plot a graph of the results shown in the table. Time should be plotted on the x-axis. Label the line you draw 'A'. **(6)**

   (ii) Sketch carefully on the graph the line that would be obtained:

   - I. if the same reaction was carried out at a temperature of 50 °C. Label this line 'B'. **(2)**
   - II. if an identical piece of calcium carbonate was reacted with only 20 cm^3 of the dilute acid. Label this graph 'C'. **(2)**

   (iii) Explain your answer to part (ii) I, in terms of collisions between the reactants. **(3)**

   (b) (i) The gradient (slope) of the line on the graph represents the rate of the reaction.

   Determine the gradient of the line at time t = 0 minutes and time t = 5 minutes. **(6)**

   (ii) Is the rate of reaction is faster at t = 0 minutes or at t = 5 minutes? Explain your answer. **(2)**

   (iii) Calculate the value of (rate at t = 0 minutes) / (rate at t = 5 minutes). **(2)**

D Equilibria

You will be expected to:

* ★ explain that some reactions are reversible and this is indicated by ⇌ in equations
* ★ describe some reversible reactions
* ★ explain the concept of dynamic equilibrium
* ★ predict the effect of changing the temperature and pressure on the equilibrium position in reversible reactions.

Reversible reactions

In some chemical reactions the products can react together to produce the original reactants. These are known as **reversible reactions.**

Examples of reversible reactions

$$\text{ammonium chloride} \underset{\text{cool}}{\overset{\text{heat}}{\rightleftharpoons}} \text{ammonia} + \text{hydrogen chloride}$$

[white solid]   [colourless gases]

$$\text{hydrated copper sulfate} \rightleftharpoons \text{anhydrous copper sulfate} + \text{water}$$

[blue solid]   [white solid]

Fig. 4d.01: Hydrated copper sulfate (left) and anhydrous copper sulfate (right)

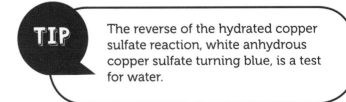

TIP The reverse of the hydrated copper sulfate reaction, white anhydrous copper sulfate turning blue, is a test for water.

Equilibrium

When a reversible reaction takes place in a **closed system** (nothing is added or removed) **equilibrium** is reached when the forward and reverse reactions occur at the same rate. At equilibrium the concentrations of the reagents and products do not change over time so long as the temperature and pressure are not changed. This can give the incorrect impression that the reaction has ceased, but in reality a **dynamic equilibrium** has been established, with reactants being converted to products at the same rate as the products react to form the reactants again.

If a reversible reaction is *exothermic* in one direction, it is *endothermic* in the opposite direction. For example, in the manufacture of ammonia the forward reaction is exothermic, and the reverse reaction is endothermic.

exothermic

$$N_2 + 3H_2 \rightleftharpoons 2NH_3$$

endothermic

The **equilibrium position** (and therefore the amounts of products in the mixture, known as the **yield**) is affected by the temperature and pressure.

Effect of temperature on equilibrium position

- An increase in temperature will move the equilibrium position in the direction of the endothermic reaction, taking in more heat, e.g. more nitrogen and hydrogen and less ammonia in the mixture.
- A decrease in temperature will move the equilibrium position in the direction of the exothermic reaction, giving out more heat, e.g. more ammonia and less nitrogen and hydrogen in the mixture.

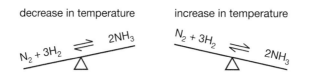

Fig. 4d.02: Equilibria and temperature

Effect of pressure on equilibrium position

- An increase in pressure will move the equilibrium position in the direction of the reaction that has the smaller number of moles.
- A decrease in pressure will move the equilibrium position in the direction of the reaction that has the larger number of moles.

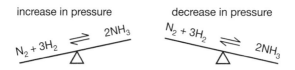

Fig. 4d.03: Equilibria and pressure

In 1884, the French chemist Le Chatelier stated the following: When a system in equilibrium is subjected to a change in conditions (e.g. temperature or pressure) the position of equilibrium shifts in such a way as to oppose the change. This is known as **Le Chatelier's Principle**.

You do not have to learn this, but you need to appreciate how commercial considerations lead to compromises in the conditions which Le Chatelier's Principle suggests might be used to obtain the greatest yield.

You should now be able to:

* explain the meaning of the symbol $\rightleftharpoons$ in chemical equations (see page 161)
* describe the effect of heat on ammonium chloride and explain why this is a reversible reaction (see page 161)
* explain the term *dynamic equilibrium* (see page 162)
* predict the effect of: (a) increasing temperature, and (b) increasing pressure, on the position of equilibrium in the following reaction:
* $4NH_3 + 3O_2 \rightleftharpoons 2N_2 + 6H_2O$ (g) $\Delta H = -1267$ kJ mol^{-1} (see page 162).

Practice questions

1. For each reaction, predict the effect of the suggested changes on the equilibrium position of the reaction and give an explanation for your choice.

    (a) Increasing pressure on the reaction: $CO(g) + 2H_2(g) \rightleftharpoons CH_3OH(g)$ **(2)**

    (b) Decreasing temperature on the reaction: $2SO_2(g) + O_2(g) \rightleftharpoons 2SO_3(g)$ (note: forward reaction is exothermic) **(2)**

    (c) Increasing temperature on the reaction: $N_2O_4(g) \rightleftharpoons 2NO_2(g)$ (note: forward reaction is endothermic) **(2)**

    (d) Decreasing pressure on the reaction: $2NH_3 \rightleftharpoons N_2 + 3H_2$ **(2)**

    (e) Decreasing temperature on the reaction: $N_2(g) + O_2(g) \rightleftharpoons 2NO(g)$ (note: forward reaction is endothermic) **(2)**

Section Five

Chemistry in society

A Extraction and uses of metal

You will be expected to:

★ explain how the reactivity of a metal is related to the method used to extract it from its ores
★ describe and explain how aluminium is extracted by electrolysis of its purified oxide in molten cryolite
★ write ionic half-equations for the reactions during aluminium extraction
★ describe and explain the main reactions in the extraction of iron in a blast furnace
★ explain the uses of aluminium and iron in terms of their properties.

The extraction of metals

Apart from a few fairly unreactive metals such as gold and silver, which are found as pure 'native' metal, most metals are found in the Earth's crust as compounds, often oxides. The metal has to be obtained by chemical **extraction**.

The most common extraction process involves reduction of the metal oxide by heating it with a reducing **agent**. Carbon is used because it is a cheap, plentiful reducing agent.

The oxides of aluminium and other reactive metals cannot be reduced by heating with carbon, so these metals are extracted by **electrolysis**.

Reactivity of metal	Examples	Extraction method
low	gold, silver	extraction not usually needed, found pure
intermediate	zinc, iron, copper	heat metal oxide with carbon
high	aluminium, sodium	electrolysis

Aluminium extraction

Aluminium is extracted from purified aluminium oxide (Al_2O_3) by electrolysis.

- The melting point of purified aluminium oxide is too high to allow it to be melted economically, so **cryolite** (Na_3AlF_6) is added to lower the melting point and to act as a solvent.
- The electrolyte of purified aluminium oxide and cryolite is kept molten at about 1000 °C.
- The graphite (carbon) cathode forms the lining of the electrolytic cell.
- The graphite (carbon) anode is suspended about 6 cm above the cathode.
- The process consumes large amounts of electricity.

An aluminium smelter needs to be:

- close to a supply of cheap electricity (e.g. hydroelectric or nuclear)
- near a deep-water port to allow easy access for the importing of the oxide
- served by good transport links for product distribution.

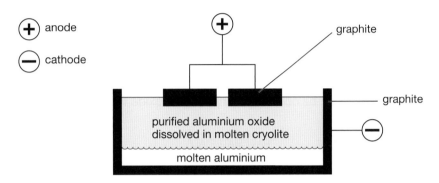

Fig. 5a.01: The electrolysis of aluminium

Reactions at the electrodes

At the cathode:

- Aluminium (molten) is formed and collects in the bottom of the cell.
- The aluminium is removed periodically without interrupting the electrolysis.

The ionic half-equation for this reaction is:

$$Al^{3+} + 3e^- \rightarrow Al$$

At the anode:

- Oxygen is formed.
- The carbon anode slowly burns away in the oxygen ($C + O_2 \rightarrow CO_2$) and must be replaced periodically.

The ionic half-equation for this reaction is:

$$2O^{2-} - 4e^- \rightarrow O_2$$

Uses of aluminium

Aluminium is a typical metal (strong, good conductor of heat and electricity) but it has a low density (about one-third that of iron). It is much more costly than iron and is used where its low density and resistance to corrosion are worth paying for, such as:

- to make aircraft because it is strong and has a low density
- in food containers and window frames because it is resistant to corrosion.

CAM

Aluminium is also used in overhead power lines because it is a good electrical conductor and has a low density.

Copper is a better electrical conductor than aluminium, but is over three times as dense and is better suited to underground cables, where weight is not an issue. To carry the same current as copper power lines, those made of aluminium must be thicker, but the lower density of aluminium means that there is still a saving in weight. To prevent the aluminium power lines from sagging under their own weight they are made with a steel core.

Iron extraction

Iron is not a very reactive metal so it can be obtained by heating its oxide with carbon in a **blast furnace**.

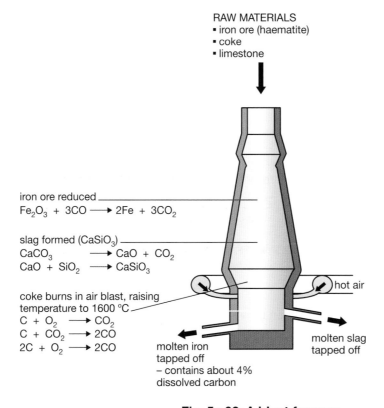

Fig. 5a.02: A blast furnace

The raw materials are:

- iron ore (haematite), which is present in sufficiently large quantities in one location to allow it to be mined commercially
- coke, to provide the carbon needed for reduction of the ore
- limestone, which reacts with acidic impurities in the ore to form a liquid **slag** that floats on top of the molten iron and is tapped off from time to time.

Hot air is added to supply the oxygen needed for combustion of the coke to produce carbon dioxide, which reacts with more carbon to produce carbon monoxide. Carbon monoxide reduces the iron oxide to iron.

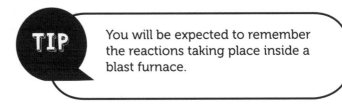

TIP — You will be expected to remember the reactions taking place inside a blast furnace.

The main reactions in a blast furnace

- Coke burns to give carbon dioxide and heat:
  coke + oxygen $\rightarrow$ carbon dioxide + heat energy
  $C(s)$ + $O_2(g)$ $\rightarrow$ $CO_2(g)$ + heat energy

- Carbon dioxide reacts with hot carbon to form carbon monoxide:
  carbon dioxide + carbon $\rightarrow$ carbon monoxide
  $CO_2(g)$ + $C(s)$ $\rightarrow$ $2CO(g)$

- Carbon monoxide reduces the iron oxide in the ore to iron:
  carbon monoxide + iron oxide $\rightarrow$ carbon dioxide + iron
  $3CO(g)$ + $Fe_2O_3(s)$ $\rightarrow$ $3CO_2(g)$ + $2Fe(l)$

Uses of iron

The iron from the blast furnace (called *pig iron*) or high carbon steel contains 4% carbon. It is very hard, but brittle. It is used to make articles such as engine blocks, manhole covers and cookware, but is too brittle to be used for much else.

Steel

Steels are mixtures of iron with carbon and other elements – these mixtures are called **alloys**.

Pig iron is converted to steel by the following process:

*   A jet of oxygen is blown over molten pig iron that has been mixed with recycled scrap iron.

*   The carbon burns off as carbon dioxide.

*   Other elements in the iron are converted to acidic oxides and these are removed by adding calcium carbonate.

The process reduces the carbon content to about 1% to give the steel known as mild steel (sometimes called low carbon steel). This is widely used because it is easy to press into the required shape, e.g. to make steel sheet for car bodies.

The addition of chromium and nickel (about 18% chromium and 8% nickel) gives stainless steel, which is hard and resists corrosion. It is widely used to make articles that have to be in contact with water, such as vessels for the manufacture of food and chemicals, kitchen knives and surgical equipment. Many car parts that used to be chromium plated are now made from stainless steel.

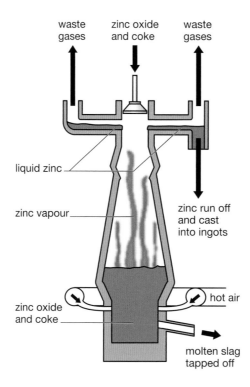

Fig. 5a.03: Zinc production in a blast furnace

The main ore of zinc is its sulfide, ZnS, known as zinc blende. Roasting zinc blende in air converts it to zinc oxide: $2ZnS + 3O_2 \rightarrow 2ZnO + 2SO_2$

The zinc oxide can be reduced by heating with carbon in the same way as iron is manufactured from its oxide. The effective reducing agent, as in the case of iron, is carbon monoxide:
$ZnO(s) + CO(g) \rightarrow Zn(g) + CO_2(g)$. The furnace used is shown in Fig. 5a.03. Zinc boils at around 900 °C and is a gas at the temperature of the furnace. It passes out of the furnace to a condenser, where it liquefies and from which it is run out and allowed to solidify into ingots.

You should now be able to:

★ explain why iron and aluminium are extracted from their ores using different processes (see page 166)

★ name the electrolyte used in the manufacture of aluminium (see page 167)

★ write down the ionic half-equations for the reactions taking place at the anode and cathode in the manufacture of aluminium (see page 167)

★ explain which electrode needs to be replaced regularly in the manufacture of aluminium, and why (see page 167)

★ explain why aluminium is an expensive metal (see page 167)

★ name the raw materials used in a blast furnace for the manufacture of iron, and explain the function of each (see page 168)

★ write equations for the blast furnace reactions in which (a) heat is produced, (b) reduction occurs, (c) slag is formed (see page 169).

Practice questions

1. a) Aluminium is prepared by electrolysis.

   (i) Write down the equation for the reaction at the anode. **(2)**

   (ii) Write down the equation for the reaction at the cathode. **(2)**

   (b) The aluminium oxide is dissolved in another substance to form a solution at 1000 °C.

   (i) Name this other substance. **(1)**

   (ii) Explain why the aluminium oxide is not simply melted and then electrolysed. **(2)**

   (iii) Explain why the aluminium oxide has to be in solution for electrolysis to occur. **(2)**

   (c) Which electrode has to be periodically replaced during the electrolysis, and for what reason? **(2)**

2. Give *three* essential features of the site of an aluminium smelter. (Give the most important first.) **(3)**

3. The low density of aluminium is important in most uses of aluminium. Give *three* uses of aluminium and a *further* property related to each use. **(6)**

4. The equation for the overall reaction that occurs during the electrolysis of aluminium oxide is
   $$2Al_2O_3(l) \rightarrow 4Al(l) + 3O_2(g).$$

   (a) (i) Calculate how many tonnes of aluminium oxide would be required to produce 100 tonnes of aluminium. [1 tonne = 1 000 000 g; M_r Al 27, O 16] **(3)**

   (ii) Calculate how many dm³ of gaseous oxygen, measured at room temperature and pressure would be formed at the same time. [1 mole of any gas occupies 24 dm³ at room temperature and pressure] **(3)**

5. Extraction of aluminium from its ore uses electricity. Iron can be extracted from its ore by heating with carbon. Gold is found occurring naturally. Explain why these three metals are obtained in different ways. **(3)**

6. The graph below shows the production of aluminium ore (bauxite), purified aluminium oxide and aluminium metal from three different countries.

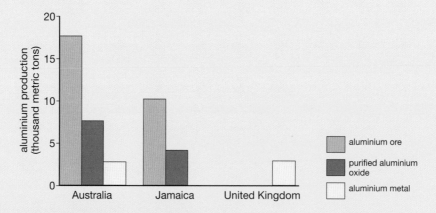

   (a) Which country has no deposits of aluminium ore? **(1)**

   (b) Suggest two reasons why Jamaica produces little aluminium metal despite mining large quantities of bauxite. **(2)**

7. (a) Name the *three* raw materials used in a blast furnace in the manufacture of iron and give the purpose of each of them. **(7)**

   (b) What is added at the base of the furnace to supply oxygen? **(2)**

8. A blast furnace is used to produce iron from iron oxide.

   (a) For this process, write equations to show:

   (i) the production of heat **(1)**
   (ii) the formation of carbon monoxide **(2)**
   (iii) reduction of the iron oxide. **(3)**

   (b) Slag is one product formed during the manufacture of iron in a blast furnace.

   (i) Name the substance added to the blast furnace to form slag. **(1)**
   (ii) Why is it economically important for the slag to be a liquid at the temperature of the blast furnace? **(2)**

9. Both solid carbon and gaseous carbon monoxide can reduce iron(III) oxide. Suggest why carbon monoxide is more effective as a reducing agent in a blast furnace. **(3)**

10. (a) (i) Explain how iron from a blast furnace is converted into steel. **(2)**

    (ii) State what is removed in the conversion of iron to steel and explain what effect this has on the properties of the steel. **(2)**

    (b) (i) Why are other metals added to steel?
    (ii) Name one metal that might be added.
    (iii) What is such a mixture of metals called? **(3)**

11. Name one iron ore. **(1)**

12. (a) (i) What is the percentage of carbon in pig iron? **(1)**
    (ii) What is the main disadvantage of pig iron as a structural material? **(1)**

    (b) (i) How is pig iron made into steel? **(2)**
    (ii) What is the carbon content of steel? **(1)**

B Crude oil

You will be expected to:

★ explain that crude oil is a mixture of hydrocarbons

★ describe the separation of crude oil into fractions using fractional distillation

★ name the main fractions from crude oil and describe their uses

★ describe the trends in boiling point and viscosity of the main fractions

★ describe how incomplete combustion of fuels may produce carbon monoxide and explain why this is toxic

★ describe how oxides of nitrogen are formed at the high temperatures reached in a car engine

★ explain how fractional distillation produces more long-chain hydrocarbons and fewer short-chain hydrocarbons than required

★ describe how catalytic cracking is used to convert long-chain alkanes to alkenes and shorter-chain alkanes.

Crude oil and fuels

CAM

The vast majority of fuels in use today, such as coal, petroleum and natural gas, are carbon-based.

Crude oil is a complex mixture of compounds, most of which contain carbon and hydrogen only and known collectively as **hydrocarbons**. As found, crude oil is a sticky black liquid that is fairly useless. However, some of the hydrocarbons in crude oil are essential to the way we live today, so they are separated from each other. Natural gas is found in the space above crude oil deposits and its main constituent is methane.

Fractional distillation

The hydrocarbons in crude oil have different boiling points and can therefore be separated by physical means using a process known as **fractional distillation**.

* The crude oil is heated and the resulting vapour passed up a tall tower known as a **fractionating column**.
* The hydrocarbons with the lowest boiling points reach the top of the tower.
* Hydrocarbons with the highest boiling points remain at the bottom of the tower.
* Those hydrocarbons with intermediate boiling points are distributed along the column according to their boiling points.

Chemistry A* Study Guide

The separated **fractions** have specific names and uses, as shown in Fig. 5b.01.

This diagram also shows the trends in boiling point and **viscosity** for the main fractions.

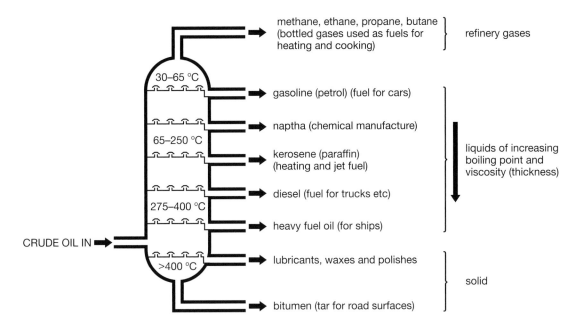

Fig. 5b.01: A fractionating column

Catalytic cracking

Fractional distillation of crude oil produces more long-chain hydrocarbons than can be used directly and fewer short-chain hydrocarbons than required. The fractions with lower boiling points, such as petrol and diesel, are more useful as fuels than the longer-chain products and so are more valuable commercially.

To increase the amount of shorter-chain products we get from crude oil, the larger-chain products are broken into shorter chains in a process called **catalytic cracking** (see Fig. 5b.02).

The vapours of the longer-chain fractions are passed over a catalyst of silica or alumina (aluminium oxide) at a temperature of 600–700 °C. **Thermal decomposition** (breakdown) takes place to form a mixture of shorter-chain alkanes and alkenes.

For example:

$$CH_3-CH_2-CH_2-CH_2-CH_2-CH_2-CH_3 \rightarrow CH_2{=}CH_2 + CH_3-CH_2-CH_2-CH_2-CH_3$$

heptane　　　　　　　　　　　　　　　ethene　　　　pentane

This example is a short-chain alkane, but the same process works for longer-chain alkanes also.

By careful control of the temperature and catalyst, the desired combination of alkene and alkane products can be obtained.

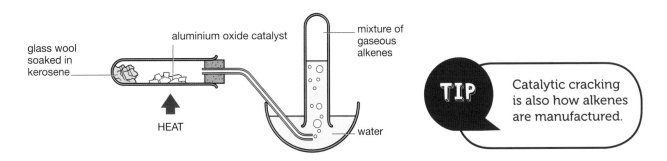

Fig. 5b.02: Laboratory demonstration of the cracking of kerosene

The combustion of fuels

The hydrocarbon fuels (natural gas, oil, etc.) burn completely in a plentiful air supply to form carbon dioxide and water plus heat energy, e.g.:

$$CH_4 + 2O_2 \rightarrow CO_2 + 2H_2O$$

Combustion is an **oxidation process**:

- carbon is oxidised to carbon dioxide
- hydrogen is oxidised to water.

Incomplete combustion

In a limited supply of air **incomplete combustion** of fuels occurs forming poisonous carbon monoxide and/or carbon.

Carbon monoxide is poisonous because it reduces the capacity of blood to carry oxygen, which affects growth and other body processes. The quantities of carbon monoxide produced from the combustion of fuels in vehicle engines are only fatal in confined spaces, but the gas can damage health in smaller amounts. As well as posing a risk to health, incomplete combustion wastes fuel and reduces engine performance. The equation for the incomplete combustion of methane is:

$$CH_4 + 1\tfrac{1}{2}O_2 \rightarrow CO + 2H_2O$$

CAM

The carbon cycle

For millions of years the amount of carbon dioxide in the Earth's atmosphere remained fairly constant. The rate at which carbon dioxide was released from natural sources such as volcanoes and vegetation fires was balanced by removal by photosynthesis and dissolving in the oceans. The main processes responsible for maintaining this balance make up the carbon cycle (see Fig. 5b.03).

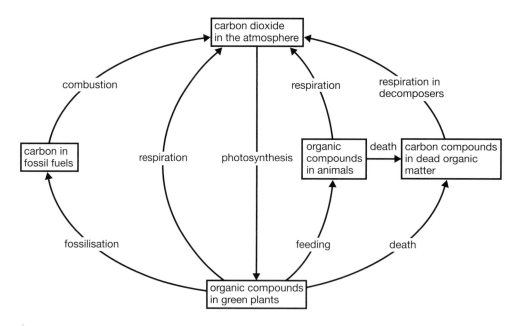

Fig. 5b.03: The carbon cycle

Industrial activity and transportation have added carbon dioxide to the atmosphere at a faster rate than natural processes can remove it, leading to an increase in the Earth's temperature, known as **global warming**.

Formation of nitrogen oxides

At the high temperatures found in a car engine, nitrogen and oxygen react to form nitrogen monoxide and nitrogen dioxide, commonly referred to together as **nitrogen oxides**.

$$N_2 + O_2 \rightarrow 2NO$$
$$2NO + O_2 \rightarrow 2NO_2$$

Catalytic converters

Nitrogen monoxide is a respiratory irritant and carbon monoxide is toxic so, to help reduce the amounts of these gases released into the air from car exhausts, catalytic converters are fitted to vehicle exhausts.

- The catalyst used is a mixture of transition metals, including platinum, rhodium and palladium.
- Carbon monoxide and unburnt hydrocarbons are oxidised in excess air added to the exhaust:
$$CO + \tfrac{1}{2}O_2 \rightarrow CO_2$$
- The oxides of nitrogen are removed by reactions with carbon monoxide.
$$CO + NO_2 \rightarrow CO_2 + NO$$
followed by: $2CO + 2NO \rightarrow 2CO_2 + N_2$

You should now be able to:

★ show how crude oil is a mixture of hydrocarbons (see page 175)

★ describe the separation of crude oil into fractions using fractional distillation (see page 175)

★ give the names and uses of the main fractions of crude oil (see page 175)

★ describe the trends in boiling point and viscosity of the main fractions (see page 175)

★ explain the relative commercial values of the various fractions (see page 175)

★ explain how and why catalytic cracking is carried out (see page 175)

★ name the products of incomplete combustion of hydrocarbons (see page 176)

★ explain why carbon monoxide is toxic (see page 176).

★ explain how the combustion of fossil fuels can harm the environment (see page 176)

★ describe the formation of oxides of nitrogen in a car engine (see page 177).

Practice questions

1. Copy and complete the passage below.

    Crude oil is a mixture of compounds containing _____ and _____ only. These compounds are called _____. The mixture can be separated by the process of _____. This process separates the mixture into _____ containing several compounds that have similar _____. **(6)**

2. Describe the effect of an increase in the number of carbon atoms on these properties of a hydrocarbon:

    (a) boiling point   (b) viscosity. **(2)**

3. (a) Name the substances formed when a hydrocarbon burns:

    (i) in a plentiful supply of air
    (ii) in a limited supply of air. **(3)**

    (b) Explain why it is important to burn fossil fuels in a plentiful supply of air. **(2)**

4. (a) Write an equation for the formation of a nitrogen oxide in a car engine. **(2)**

    (b) Explain why this reaction occurs in a car engine, but not at normal laboratory temperatures. **(3)**

5. The apparatus shown below can be used to demonstrate catalytic cracking in the laboratory. The glass wool must be heated from time to time if the experiment is to work.

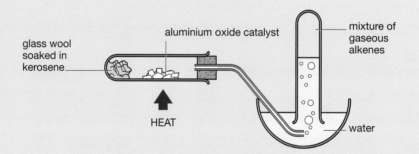

Paraffin is a mixture of hydrocarbons. One of the components in the mixture has the formula $C_{10}H_{22}$

(a) (i) To what homologous series does $C_{10}H_{22}$ belong? Explain your answer. **(2)**
    ii) State and explain what you would expect to see if bromine water was added to $C_{10}H_{22}$ and the mixture shaken. **(3)**

(b) (i) One of the products obtained when $C_{10}H_{22}$ is cracked is butene, which has the formula C_4H_8
         Write an equation for this cracking reaction. **(2)**
    (ii) State one commercial reason for the catalytic cracking. **(1)**

(c) (i) Suggest why it is necessary to heat the glass wool from time to time. **(2)**
    (ii) During the cracking experiment an oily film appears on the water surface. Suggest how and why it forms. **(2)**

Chemistry A* Study Guide

C Synthetic polymers

You will be expected to:

★ define the terms *addition polymer* and *condensation polymer* and give examples
★ describe how addition polymers are made by joining together many monomers
★ draw the repeat unit of the addition polymers poly(ethene), poly(propene) and poly(chloroethene)
★ deduce the structure of a monomer from the repeat unit of an addition polymer
★ describe how, during condensation polymerisation, a small molecule (e.g. water or hydrogen chloride) is formed
★ name the types of monomers used in the manufacture of nylon
★ draw the structure of nylon as a block diagram.

Addition polymerisation

A **polymer** is large molecule made of many identical molecules called **monomers**. **Addition polymers** are formed from only one kind of monomer.

For example, alkenes form addition polymers when the C=C double bonds of the alkene monomer are broken and the monomers join together.

$CH_2=CH_2$ + $CH_2=CH_2$ + $CH_2=CH_2$ + $CH_2=CH_2$ + $CH_2=CH_2$ + $CH_2=CH_2$ + ... →

  $-CH_2-CH_2-CH_2-CH_2-CH_2-CH_2-CH_2-CH_2-CH_2-CH_2-CH_2-CH_2-...$

The structures of some other important addition polymers, and their corresponding monomers are shown in the table.

Monomer	Polymer
H H | | C═C | | H H ethene	H H H H | | | | —C—C—C—C— | | | | H H H H poly(ethene)
CH_3 H | | C═C | | H H propene	CH_3 H CH_3 H | | | | —C—C—C—C— | | | | H H H H poly(propene)
Cl H | | C═C | | H H chloroethene	Cl H Cl H | | | | —C—C—C—C— | | | | H H H H poly(chloroethene)

Uses of synthetic addition polymers

Polymer	Properties	Uses
Poly(ethene) [polythene]	Resistant to heat, water and chemicals. Flexible. Poor mechanical strength	Film in packaging and for making containers. Electrical insulation. High-density PE is used for pipes and buckets etc.
Poly(propene) [polypropylene]	Hard and very strong. High melting point	Household goods and toys Sterilisable medical equipment Tubes and pipes, fibres for ropes
Poly(chloroethene) [PVC - polyvinyl chloride]	High chemical stability, comparatively cheap	Window frames, guttering, containers, floor coverings, waterproof clothing, electrical insulation
Poly(tetrafluoroethene) [Teflon]	Very high chemical stability, low coefficient of friction, impermeable to most substances	Bearings, seals and gaskets, linings for chemical apparatus, non-stick coatings
Poly(phenylethene) [polystyrene]	Low cost, easy to process, excellent acid and alkali resistance but damaged by organic solvents	Packaging, insulation, fillings for furniture (foam)

Condensation polymerisation

- Condensation polymers are formed from at least two different kinds of monomers.
- The two monomers must both have two functional groups that are able to react with each other.
- As each linkage between monomers is formed, a small molecule (e.g. water or hydrogen chloride) is released.

For example, dicarboxylic acids and dialcohols can react together forming an ester at each end, with the loss of a molecule of water as each ester linkage is formed:

Fig. 5c.01: Formation of a condensation polymer

TIP

The lozenge and triangular shapes shown in Fig. 5c.02 represent organic chains, and the simplified diagram is known as a **block diagram** of the polymer. The resulting polymer is an example of a **polyester** because it is made up of a large number of ester linkages.

Although knowledge of polyesters is not required, they have been used as an example here because the formation of simple esters is familiar to you.

Nylon

Nylon is a **polyamide** formed from the two monomers shown in Fig. 5c.02. In this reaction a molecule of hydrogen chloride (HCl) is formed every time two monomers react.

The monomers can be represented more simply in a block diagram

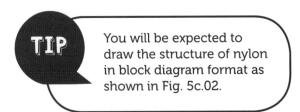

Fig. 5c.02: Block diagram to show the formation of nylon

The amide linkage (NH–C=O) gives its name to this type of polymer.

> **TIP**
> You will be expected to draw the structure of nylon in block diagram format as shown in Fig. 5c.02.

CAM

Terylene is a **polyester** formed from the two monomers shown below.

benzene-1,4-dicarboxylic acid   HOOC—⬡—COOH (shown as) HOOC—▮—COOH

ethylene glycol (shown as) HO—▯—OH

A molecule of water is formed every time two monomers react.

The resulting polymer can be represented as follows:

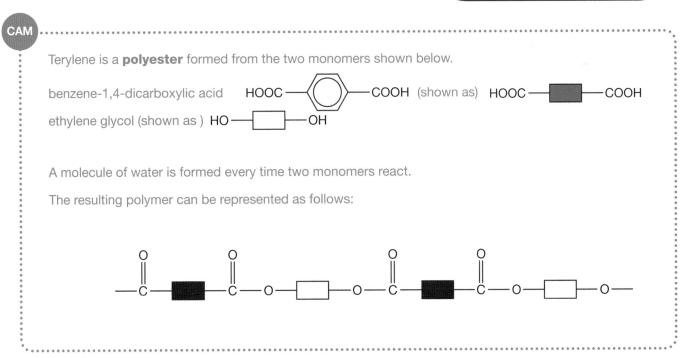

Uses of synthetic condensation polymers

Polymer	Uses
Nylon	Woven into fibres to make fabric for clothing, ropes and bristles for brushes Moulded to make components for electrical equipment and domestic appliances Tyre manufacture
Terylene	Woven into fibres to make fabric for clothing Moulded into recyclable plastic containers

Expanded polystyrene, which is widely used to package take-away meals and goods such as refrigerators and television sets, is the main cause of pollution problems by discarded polymers. It takes a very long time to disappear because decomposers (insects, bacteria, fungi, moulds etc.) cannot break it down biologically. Many plastics used for packaging now have weak links in the polymer chain, which enable them to be broken up into smaller chains on exposure to light and oxygen, making it easier for decomposers to biodegrade them. Any discarded plastic becomes a visual pollutant, and can cause internal damage if eaten by animals.

Natural macromolecules

The main constituents of food are proteins, fats and carbohydrates.

Proteins

Proteins contain the same amide linkages as are found in nylon (see page 181). These linkages are formed from amino acids, which have the general structure shown in Fig. 5c.03, where R represents an alkyl group, such as CH_3- or $-CH(CH_3)_2$

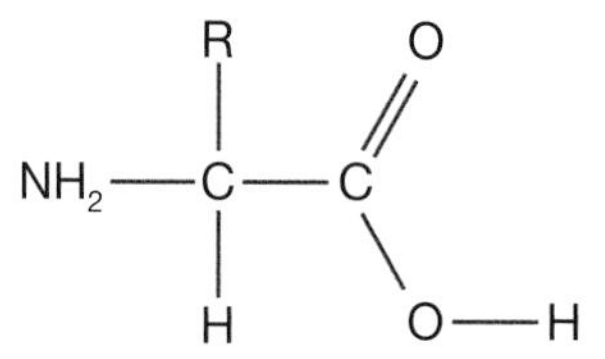

Fig. 5c.03: An amino acid

The –COOH group of one amino acid can react with the –NH_2 of another to form a **peptide** and water. The reaction is an example of **condensation polymerisation** (see Fig. 5c.04).

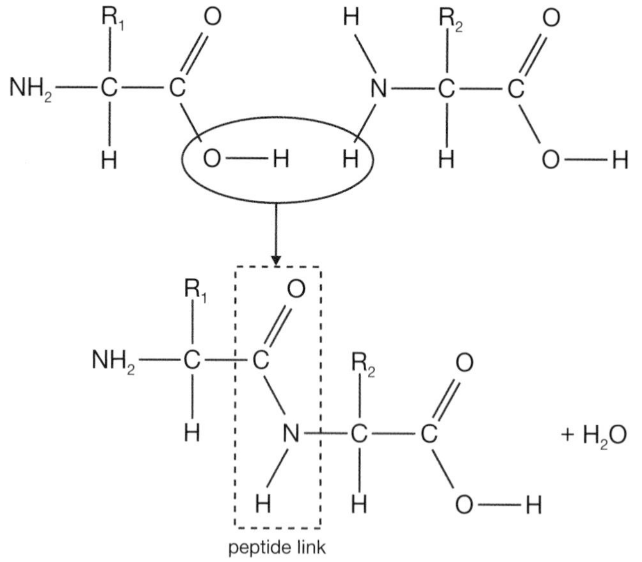

peptide link

Fig. 5c.04: Peptide bond formation

A general protein can be represented as shown in Fig. 5c.05.

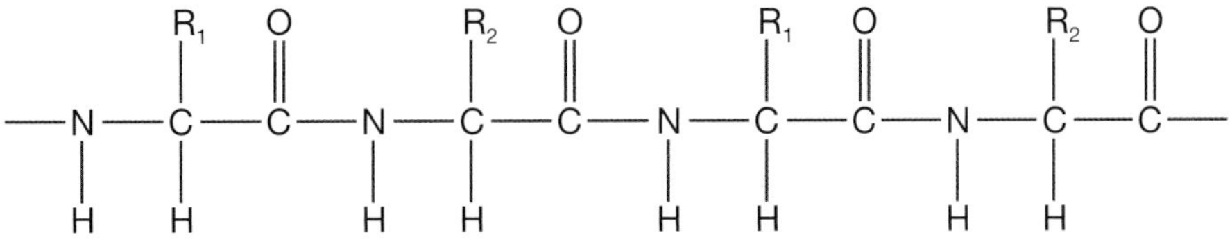

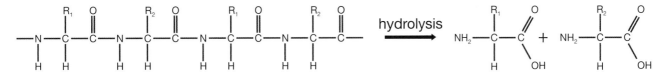

Fig. 5c.05: A general protein

Proteins in food can be **hydrolysed** by the strong hydrochloric acid in the stomach to give amino acids again (see Fig. 5c.06). The body then uses these amino acids to synthesise the specific proteins it requires to build structural materials, such as cartilage, skin, tendons and blood vessels.

Fig. 5c.06: Hydrolysis of proteins in food

Fats

Fats are esters of glycerol (propane-1,2,3-triol) containing ester linkages like those found in terylene, but involving different units. The carboxylic acids contained in fats have long hydrocarbon chains containing 15–25 carbon atoms.

The structure of a typical fat and the products of its hydrolysis with aqueous sodium hydroxide are shown in Fig. 5c.07.

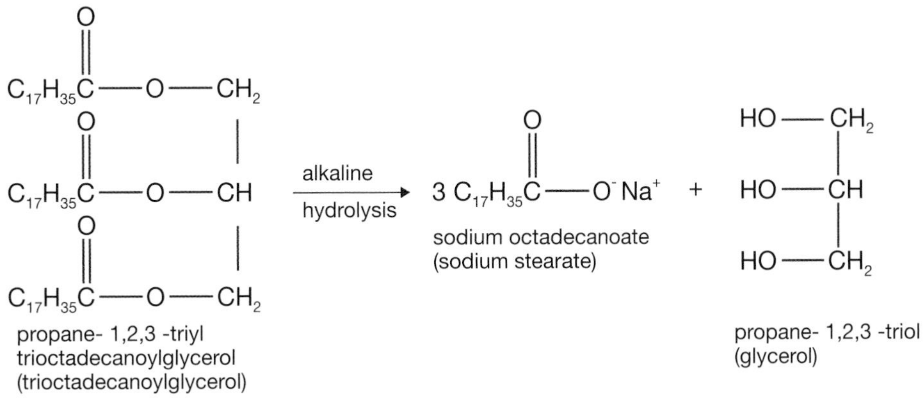

propane- 1,2,3 -triyl
trioctadecanoylglycerol
(trioctadecanoylglycerol)

sodium octadecanoate
(sodium stearate)

propane- 1,2,3 -triol
(glycerol)

Fig. 5c.07: Hydrolysis of a typical fat with aqueous sodium hydroxide

The resulting sodium salts of the long-chain carboxylic acid acids are **soaps**. Soaps contain a hydrocarbon chain (soluble in grease) and a polar carboxylic acid group (soluble in water) (see Fig. 5c.08).

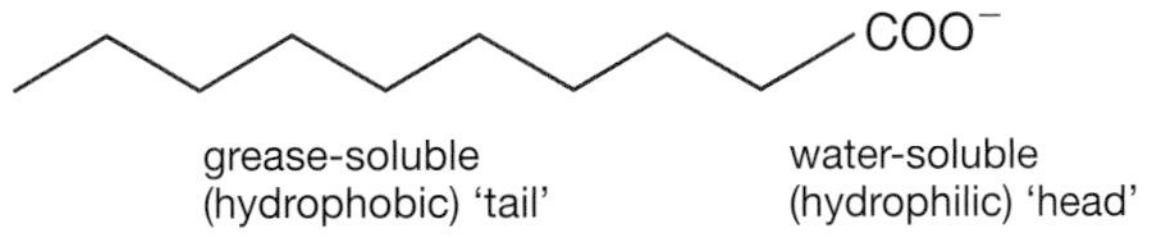

grease-soluble
(hydrophobic) 'tail'

water-soluble
(hydrophilic) 'head'

Fig. 5c.08: A soap with its hydrocarbon chain and polar carboxylic acid group

The hydrocarbon part dissolves in grease, which is lifted from soiled surfaces by the attraction of the water for the polar COO^- group.

Carbohydrates

Carbohydrates consist of large numbers of sugar units, joined as a result of condensation polymerisation.
Fig. 5c.09 shows how two sugar units link together. Further condensation reactions can occur at each end to form a longer polymer chain.

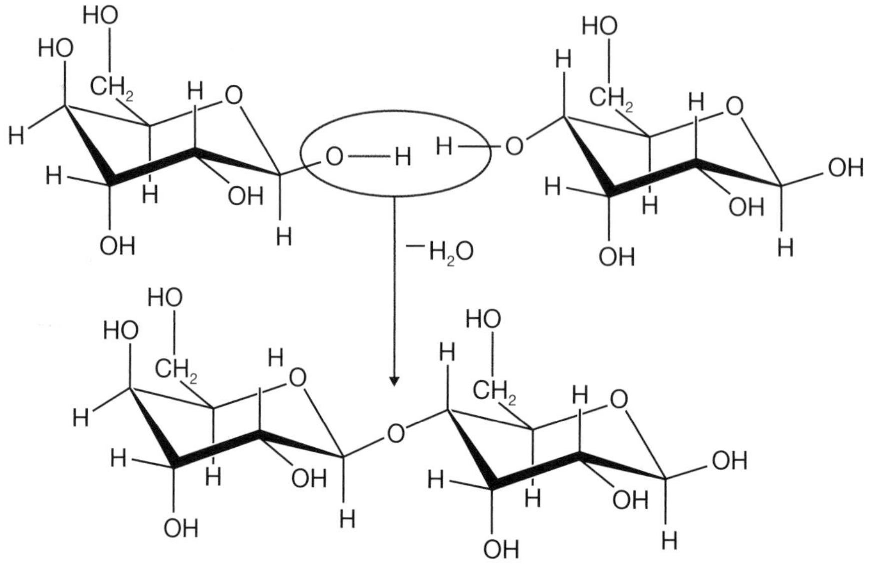

Fig. 5c.09: Two sugar units linked together

The complicated ring structures can be simplified as block diagrams, and you should recognise diagrams like Fig. 5c.10 as representing carbohydrates.

Fig. 5c.10: Block diagrams of carbohydrate

Complex carbohydrates, such as starch, are hydrolysed by acid to give simple sugars (see Fig. 5c.11):

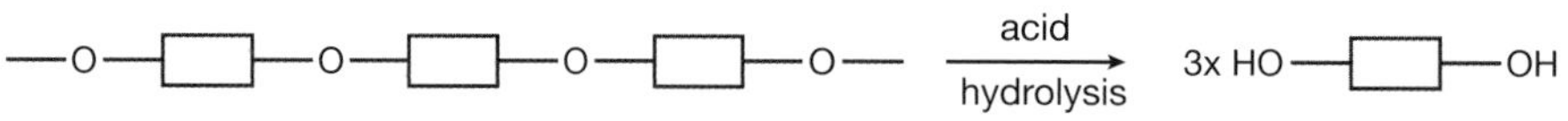

Fig. 5c.11: A complex carbohydrate hydrolysed by acid to give simple sugars

where   HO —[ ]— OH   represents

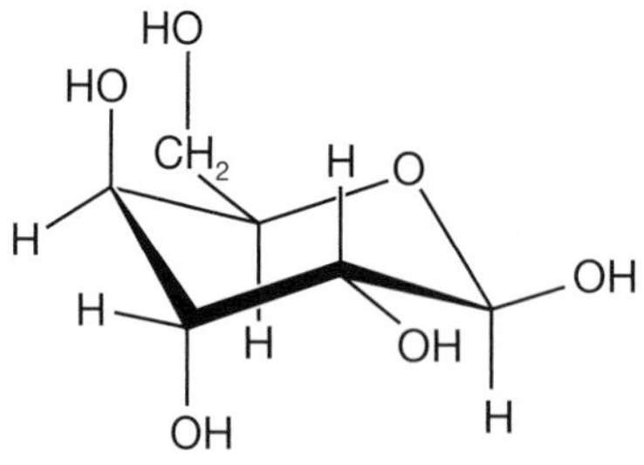

Chromatography and the hydrolysis of carbohydrates and proteins

Carbohydrates and proteins are complex molecules which, when hydrolysed, yield mixtures of sugars and amino acids, respectively. Chromatography can be used to identify and separate the individual constituents of such mixtures. For identification purposes paper chromatography of the kind described in Section 1b can be used. Because amino acids and sugars are colourless a **locating agent** must be used to make the spots on the paper visible. The various components can be identified by comparing their R$_f$ values (see page 11) with those of known samples of amino acids and sugars.

Separating the mixtures to allow the individual sugars or amino acids to be obtained is more complicated. The principal method used is high-performance liquid chromatography. Although details of this technique are not required for the examination, you may wish to investigate it further for your own interest.

You should now be able to:

★ define the terms *monomer* and *polymer* (see page 179)
★ explain the basic differences between addition polymers and condensation polymers (see page 180)
★ draw the repeat units of poly(ethene), poly(propene) and poly(chloroethene) (see page 179)
★ write down the structure of an addition polymer from the repeat unit of its monomer (see page 179)
★ give an example of a condensation polymer (see page 180)
★ explain which types of monomers are used in the manufacture of nylon (see page 181)
★ draw the structure of nylon as a block diagram (see page 181).

Practice questions

1. a) Write down *three* repeat units of the polymer formed from each alkene below.

    (i) ethene $\qquad$ $CH_2=CH_2$ $\qquad$ poly(ethene) **(1)**

    (ii) chloroethene $\qquad$ $CHCl=CH_2$ $\qquad$ poly(chloroethene) **(2)**

    (b) Write down the structure of the monomer used to make each of the polymers shown below. (Two repeat units have been shown for each.)

    (i) $-CHCl-CCl_2-CHCl-CCl_2-$

    (ii) $-CHCH_3-CHCH_3-CHCH_3-CHCH_3-$ **(2)**

CAM 2. State one use for each of the following polymers.

    (a) poly(ethene) $\qquad$ (b) poly(chloroethene) $\qquad$ (c) poly(propene) **(3)**

3. Discarded plastics can create environmental problems. State the main source of these plastics and explain why they cause environmental problems. **(3)**

4. (a) Describe *two* differences between an addition polymer, such as poly(ethene), and a condensation polymer, such as nylon, in terms of how each is formed from their monomer(s). **(4)**

CAM (b) Give one use each for nylon and terylene. **(2)**

D The manufacture of some important chemicals

Ammonia manufacture

The raw materials for manufacturing ammonia are:

- nitrogen extracted from the air
- hydrogen, from natural gas, or the cracking of hydrocarbons.

The **Haber Process** is used to manufacture ammonia.

The equation for the reaction is:

$$N_2 + 3H_2 \rightleftharpoons 2NH_3 + \text{heat energy}$$

The reaction is reversible. The reaction to form ammonia is **exothermic** and involves a *reduction in volume*. So the ideal reaction conditions for producing as much ammonia as possible (the maximum **yield**) would be:

- highest possible pressure (the formation of ammonia is favoured because that reduces the pressure)
- lowest possible temperature (the formation of ammonia is favoured because that increases the temperature).

But high pressures are costly to produce and low temperatures reduce the rate of reaction. So, in the manufacturing process a *compromise* must be reached and the actual reaction conditions are:

- temperature of 450 °C
- compressed to 250 atmospheres
- the gases are passed over a catalyst of iron, finely divided to increase the contact between the reactant gases and the catalyst surface.

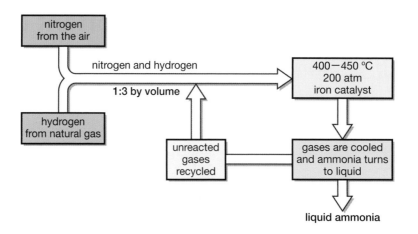

Fig. 5d.01: The Haber process

The catalyst has no effect on the position of equilibrium, but allows equilibrium to be established more rapidly. The catalyst costs money but it allows the manufacturer to obtain the same yield at a lower temperature, thus saving more money than the expenditure on the catalyst.

About 15% of the gases are converted to ammonia at each passage over the catalyst so the unreacted nitrogen and hydrogen are re-cycled over the catalyst several times to increase the overall yield.

The ammonia formed in each pass over the catalyst is condensed to a liquid (boiling point –33 °C) and separated from the unreacted gases.

Ammonia has important uses:

- to make fertilisers such as ammonium nitrate and sulfate
- as the starting material for nitric acid manufacture.

Sulfuric acid

Sulfuric acid is manufactured using the **contact process**.

The raw materials for the process are:

- oxygen from air
- sulfur dioxide, produced by burning sulfur in air
- water.

Molten sulfur is sprayed into a furnace and burnt in a blast of dry, dust-free air to form sulfur dioxide. The temperature rises to over 1000 °C.

The mixture of sulfur dioxide and excess air is then cooled (by passing it through boilers to raise steam) to about 450 °C and passed over a catalyst of vanadium pentoxide to form sulfur trioxide.

The reaction is:

$$2SO_2(g) + O_2(g) \rightleftharpoons 2SO_3(g) + \textit{heat energy}$$

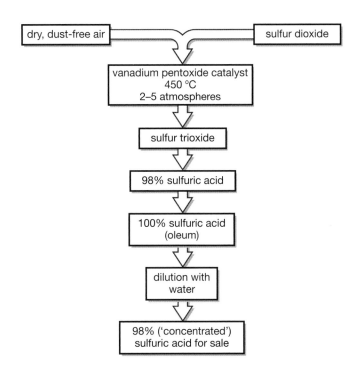

Fig. 5d.02: The contact process

This reaction is reversible, is *exothermic* in the formation of sulfur trioxide and involves a (slight) *reduction in volume*. So the maximum yield of sulfur trioxide would be obtained at:

* low temperature, because this favours the reaction producing heat so forming more sulfur trioxide
* high pressure, because this favours the reaction resulting in a reduction in volume so producing more sulfur trioxide.

But:

* at low temperature the reaction is too slow so a compromise temperature of 450 °C is used: high enough to give an acceptable yield of sulfur trioxide rapidly
* the small reduction in volume that occurs (3 moles to 2 moles) means that the yield increases very little as the pressure is increased and not enough to offset the cost of the necessary pumps.

The actual operating conditions used are:

* temperature of 450 °C
* pressure of 2–5 atmospheres
* catalyst of vanadium pentoxide, V_2O_5, finely divided to increase the contact between the reactant gases and the catalyst surface.

The catalyst does not affect the eventual yield of sulfur trioxide, but only *increases the rate at which it is formed*.

Under optimum conditions a 98% yield of sulfur trioxide is obtained by passing the reactant gases through several catalytic converters in series.

Conversion of sulfur trioxide to sulfuric acid

Sulfur trioxide reacts very exothermically with water to form sulfuric acid:

$$SO_3 + H_2O \rightarrow H_2SO_4$$

However, this reaction produces too much heat and a dense mist of sulfuric acid forms, which cannot be condensed. To overcome this problem the sulfur trioxide is absorbed in 98% sulfuric acid (the other 2% being water), which is maintained at 98% by the simultaneous addition of water.

Commercial 'concentrated' sulfuric acid is the 98% product.

Sulfuric acid is used in the manufacture of:

* fertilisers
* detergents
* paints.

Manufacturing sodium hydroxide and chlorine

Sodium hydroxide and chlorine are manufactured commercially in a **diaphragm cell** in which brine (saturated aqueous sodium chloride) is electrolysed.

* Sodium hydroxide solution forms in the region around the cathode (negative electrode).
* Chlorine gas is formed at the anode.

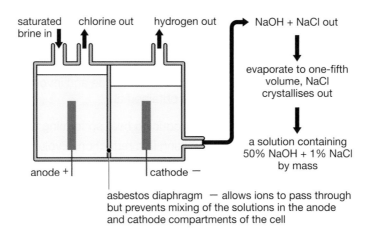

Fig. 5d.03: A diaphragm cell

Ionic half-equations

The ions originally present in NaCl(aq) are Na^+, Cl^-, H^+ and OH^-.

* Cl^- is removed during electrolysis as $Cl_2(g)$ at the anode.
* H^+ is removed during electrolysis as $H_2(g)$ at the cathode.
* Na^+ and OH^- ions remain in solution as NaOH(aq).

Uses of sodium hydroxide

Sodium hydroxide is used in the manufacture of:

* bleach
* paper
* soap.

Uses of chlorine

Chlorine produced in this process is also important commercially, and is used to:

* sterilise water (drinking water supplies and swimming pools) because in solution it kills bacteria
* manufacture bleach
* manufacture hydrochloric acid.

You should now be able to:

★ name the raw materials and reactants in the Haber Process for the manufacture of ammonia (see page 187)

★ state the temperature, pressure and catalyst used in the Haber Process (see page 187)

★ explain why a lower temperature is not used in the Haber Process, even though this would increase the yield of ammonia (see page 187)

★ state two uses of ammonia (see page 188)

★ name the raw materials and reactants in the Contact Process for the manufacture of sulfuric acid (see page 188)

★ explain why the pressure used in the Contact Process is close to atmospheric pressure (see page 189)

★ state two uses of sulfuric acid (see page 190)

★ name the electrolyte, the product formed at the anode and the product formed at the cathode during the manufacture of sodium hydroxide in a diaphragm cell (see page 190)

★ write the ionic half-equations for the cathode and anode reactions in the manufacture of sodium hydroxide in a diaphragm cell (see page 190)

★ state two uses for the product formed at the anode in the manufacture of sodium hydroxide by electrolysis (see page 191)

★ state two uses of sodium hydroxide (see page 190).

Practice questions

1. The Haber Process is used to manufacture ammonia. Nitrogen gas and hydrogen gas react with each other in the presence of a catalyst.

    (a) Name the catalyst used in the Haber Process. **(1)**

    (b) The equation for the reaction is: $N_2(g) + 3H_2(g) \rightleftharpoons 2NH_3(g)$. The formation of ammonia is exothermic.

       (i) This equation represents a reversible reaction in which reactants and products are in dynamic equilibrium. Explain what is meant by the term *dynamic equilibrium*. **(2)**

       (ii) State and explain the effect of each of the following on the yield of ammonia:

          (I) an increase in pressure **(3)**

          (II) a decrease in temperature. **(3)**

       (iii) Write down the temperature and pressure actually used in the Haber Process. **(2)**

       (iv) Considering your answer to (ii), explain why these values of temperature and pressure are used. **(4)**

2. Sulfuric acid is manufactured using the Contact Process.

    (a) List the raw materials for this process. **(3)**

    (b) In this process, the reactants (sulfur dioxide and oxygen) are in equilibrium with the product (sulfur trioxide). The formation of sulfur trioxide is exothermic.

    $$2SO_2(g) + O_2(g) \rightleftharpoons 2SO_3(g)$$

       (i) If the system is in equilibrium, what must be true concerning the amounts of the three substances present? (1)

       (ii) How do the rates of the forward and back reactions compare at equilibrium? **(1)**

    (c) (i) What temperature is used in the Contact Process? **(1)**

       (ii) Explain why this temperature is used, rather than a lower temperature. **(3)**

       (iii) What pressure is used in the Contact Process? **(1)**

       (iv) Explain why a higher pressure is not used, even though the yield would be improved. **(2)**

       (v) Name the catalyst used in the Contact Process, and explain why, despite its high cost, its use makes economic sense. **(4)**

    (d) Give *two* uses for sulfuric acid. **(2)**

    (e) Sulfur trioxide is not added directly to water to make sulfuric acid. Explain why not and state what is done instead to make the concentrated acid. **(4)**

3. In the Contact Process for the manufacture of sulfuric acid, the percentage conversion of sulfur dioxide varies with the temperature. This is shown in the graph below.

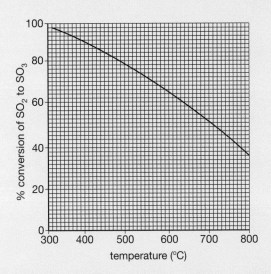

(a) Use the graph to find the percentage conversion at a temperature of 580 °C. **(1)**

(b) Calculate the mass of sulfur trioxide formed when 300 tonnes of sulfur dioxide is converted at 580 °C. [M_r S = 32; O = 16; 1 tonne = 10^6 g] **(5)**

4. Sodium hydroxide is made by electrolysing brine in a diaphragm cell.

(a) Write labels for the numbers on the diagram of a diaphragm cell below. **(8)**

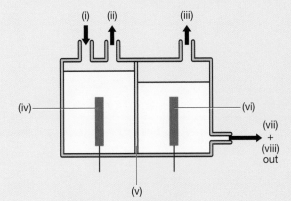

(b) Write the ionic half-equation for the formation of the gas:

   (i) at the cathode   (ii) at the anode. **(4)**

(c) Explain why the solution near the cathode becomes alkaline during the electrolysis. **(3)**

(d) Give *two* uses each for:

   (i) sodium hydroxide   (ii) chlorine. **(4)**

(e) In a particular electrolysis, 20 moles of chlorine gas were formed at the anode. How many moles of hydroxide ion would be formed at the cathode at the same time? **(3)**

Answers

1 Principle of chemistry

A States of matter

1. (a) The terms *melting, evaporation/boiling, freezing/solidifying* and *condensing* should be placed correctly as in the diagram on page 3. **(4)**

    (b) Particles in the solid are vibrating **(1)**

    heating increases the average energy of the particles and they vibrate more **(1)**

    eventually the particles have sufficient energy to overcome the forces between them so they move more freely **(1)**

2. (a) The particles in the petrol are in constant, random motion **(1)**

    they diffuse / mix with the air **(1)**

    and, given sufficient time, can travel long distances **(1)**

    (b) The particles in a liquid are in contact **(1)**

    moving them together, even to a small extent, leads to strong repulsions between the particles **(1)**

    particles in gases are far apart and can be compressed a great deal until they become close enough to each other to resist compression **(1)**

    (c) The particles in a solid vibrate about their fixed positions **(1)**

    as the temperature increases, these vibrations become greater in amplitude **(1)**

    so the particles take up more space, and the solid expands **(1)**

3. **(9)**

Property	Solid	Liquid	Gas
density	decreases ⟶		
compressibility	increases ⟶		
ease of flow	increases ⟶		
ability to maintain shape	good	poor	poor
ability to maintain volume	good	good	poor

4. When a solid is heated it **melts (1)** turning into a **liquid (1)**. Further heating leads to formation of a **gas (1)**, which returns to a liquid during the process of **condensation (1)**.

    The particles in a gas are **further (1)** apart than those in liquid. In a solid the particles cannot move from place to place, but are able to **vibrate (1)**.

B Atoms

1. (a) Formed from atoms of only one kind **(1)** having the specific properties of that element **(1)**

   (b) Two or more atoms of same or different kinds **(1)** chemically bonded together **(1)**

   (c) Two or more substances **(1)** together in the same place but not chemically bonded together **(1)**

2. B, A, G, E, D, C, F **(5)**

3. **(15)**

Separation method	Property used	Used to separate	Example
filtration	particle size	solids from liquids / solutions	sand from seawater
crystallisation	solubility	solids from solutions	salt from seawater
distillation	boiling point	liquids from solutions	water from seawater
fractional distillation	boiling point	liquids from a mixture of liquids	ethanol from aqueous ethanol
paper chromatography	speed of travel along chromatogram / retention by third substance e.g. paper	coloured substances	food colouring

4. (a) Mark a starting line on paper in pencil **(1)**

   place spots of a solution of each substance on starting line **(1)**

   dip lower edge of paper into solvent **(1)**

   leave until solvent just reaches top of paper **(1)**

   remove the paper and dry it **(1)**

   (b) (i) Pencil / graphite is insoluble **(1)**

   but ball-point pen ink would move up the paper (and spoil the experiment) **(1)**

   (ii) No **(1)**

   they have different numbers of spots / colours **(1)**

   they are in different places on the paper **(1)**

CAM (c) Correct measurement of distance travelled by solvent **(1)**

   correct measurement of distance travelled by spot 'X' **(1)**

   correct calculation of R_f value **(1)**

5. (a) Particles in both liquids are in constant random motion **(1)**

   so particles of one liquid can move between particles of the other liquid **(1)**

   eventually leading to complete mixing of the two liquids **(1)**

   (b) increasing the temperature increases the average energy of the particles **(1)**

   so they move more rapidly **(1)**

   increasing the rate of diffusion **(1)**

C Atomic structure

1. (a) Smallest particle showing the properties of a given element **(1)**

   (b) An atom which has lost or gained electron(s) **(1)**

2. **(3)**

Particle	Charge	Relative mass
proton (p)	+1	1
neutron (n)	0	1
electron (e)	−1	1/1840

3. (a) (i) nucleus **(1)**

   (ii) proton **(1)**
   neutron **(1)**

   (iii) In shells orbiting the nucleus **(1)**
   held by electrostatic attraction **(1)**

4. 10 **(1)**

   because the positive charges of the protons must be balanced by the negative charges of the electrons / because the atom overall must be neutral **(1)**

5. 1st 2, 2nd 8, 3rd 8 **(2)**

6. (a) (i) 2   (ii) 8 **(2)**

   (b) Both have a full outer shell of electrons **(1)**

7. (a) (i) 2.8.1 (ii) 2.8 **(2)**

   (b) A sodium ions has a full outer shell / inert gas / neon structure **(1)** this is very stable **(1)**

8. O^{2-} **(1)** this ion has a full outer shell inert gas / neon structure **(1)** this is very stable **(1)**

9. (a) (i) mass / nucleon (number) **(1)** proton (number) **(1)**

   (ii) 20 for all   3×**(1)**

   (b) **(15)**

ion	protons	neutrons	electrons
$^{1}_{1}H^{+}$	1	0	0
$^{9}_{4}Be^{2+}$	4	5	2
$^{56}_{26}Fe^{3+}$	26	30	23
$^{127}_{53}I^{-}$	53	74	54
$^{79}_{34}Se^{-}$	34	45	35

10. (a) Atoms with the same proton number **(1)** but different mass/nucleon numbers **(1)**

    (b) the mass of an atom **(1)** relative to 1/12 the mass of an atom of carbon-12 **(1)**

    (c) $(0.9 \times 20) + (0.1 \times 22) = 18 + 2.2 = 20.2$ **(3)**

    (d) Identical chemical properties **(1)**
    chemical reactions involve only the electrons **(1)**
    electronic structures of the isotopes are the same **(1)**

D Relative formula masses and molar volumes of gases

1.   moles hydrogen formed = 3.0 ÷ 24.0 = 0.125 mol **(1)**

from the equation 1.0 mol of iron forms 1.0 mol of hydrogen **(1)**

so 7.0g of iron = 0.125 mol **(1)**

mass of 1.0 mol of iron = 7.0 ÷ 0.125 = 56 **(1)**

2.   (a) 1 mole Fe_2O_3 = (2 × 56) + (3 × 16) = 160 g. This gives 2 × 56 = 112 g Fe.

   7.0 g of Fe would be formed when (160 ÷ 112) × 7 g of Fe_2O_3 was reduced = 10.0 g **(3)**

   (b) 10.0 g of Fe_2O_3 = 10 ÷ 160 = 0.0625 moles. 1 mole of Fe_2O_3 needs 3 moles of C to reduce it.

   Moles C = 3 × 0.0625 = 0.1875 mole. Mass C = 0.1875 × 12 = 2.25 g **(2)**

   (c) Moles CO formed = moles C used = 0.1875 (see (b) above)

   Volume of CO formed = moles CO formed × 24 dm^3 = 0.1875 × 24 = 4.5 dm^3 at room temperature and pressure **(3)**

   (d) From the equation the volume of O_2 is half the volume of CO = 4.5 ÷ 2 = 2.25 dm^3 **(2)**

3.   (a) (i)   From the equation: $4LiNO_3(s) \rightarrow 4NO_2(g)$, so 1 mole $LiNO_3$ gives 1 mole NO_2 **(1)**

   (ii) 7 + 14 + 3(16) = 69 **(1)**

   (iii) 1.38 ÷ 69 = 0.02 **(1)**

   (iv) 0.02 moles $LiNO_3$ give 0.02 moles NO_2 (from answer to (i))

   M_r NO_2 = 14 + 2(16) = 46

   0.02 moles NO_2 has a mass of 0.02 × 46 g = 0.92 g **(1)**

   (v) 1 mole of NO_2 occupies 24 dm^3 at RTP

   so 0.02 moles NO_2 occupies 0.02 × 24 dm^3 = 0.48 dm^3 or 480 cm^3 **(1)**

   (vi) From the equation: moles O_2 = ¼ × moles NO_2 so volume of O_2 = ¼ × volume NO_2 = 120 cm^3

   Total volume of gas = volume NO_2 + volume O_2 = 480 + 120 cm^3 = 600 cm^3 **(1)**

   (vii) From the equation: $4LiNO_3(s) \rightarrow 2Li_2O(s)$ so each mole of $LiNO_3$ gives half a mole of Li_2O

   M_r Li_2O = 2(7) + 16 = 30

   We started with 0.02 moles $LiNO_3$ so will be left with 0.5 × 0.02 = 0.01 moles Li_2O

   Mass of 0.01 moles Li_2O = 0.01 × 30 = 0.30g Li_2O **(2)**

   (b) We have 0.01 moles of Li_2O in 250 cm^3 of water, but each Li_2O gives 2 moles of LiOH

   Moles hydroxide ions in 250 cm^3 = 2 × 0.01 = 0.02 moles

   In 1.0 cm^3 there will be 0.02 ÷ 250 moles

   So in 1000 cm^3 (1.0 dm^3) there will be 1000(0.02 ÷ 250) = 0.08 moles **(2)**

E Chemical formulae and chemical equations

1. (a) Li_2O (b) $CaBr_2$ (c) Al_2O_3 (d) $Ca(OH)_2$ (e) $(NH_4)_2SO_4$ (f) $Zn(NO_3)_2$ (g) $Al(OH)_3$ (h) Ag_2S (i) FeF_3 (j) $Al_2(SO_4)_3$
$10\times$ **(1)**

2. (a) $CaCO_3 + 2HCl \rightarrow CaCl_2 + CO_2 + H_2O$ **(2)**

   (b) $2NaOH + H_2SO_4 \rightarrow Na_2SO_4 + 2H_2O$ **(2)**

   (c) $2K + 2H_2O \rightarrow 2KOH + H_2$ **(2)**

   (d) $4LiNO_3 \rightarrow 2Li_2O + 4NO_2 + O_2$ **(2)**

   (e) $Ca(OH)_2 + 2HNO_3 \rightarrow Ca(NO_3)_2 + 2H_2O$ **(2)**

   (f) $3CuSO_4 + 2Al \rightarrow Al_2(SO_4)_3 + 3Cu$ **(2)**

   (g) $2Al + Fe_2O_3 \rightarrow Al_2O_3 + 2Fe$ **(2)**

   (h) $4NH_3 + 5O_2 \rightarrow 4NO + 6H_2O$ **(2)**

   (i) $2NH_4Cl + Ca(OH)_2 \rightarrow 2NH_3 + CaCl_2 + 2H_2O$ **(2)**

   (j) $2Ca(NO_3)_2 \rightarrow 2CaO + 4NO_2 + O_2$ **(2)**

3. (a) $NaCl$ (b) $NaOH$ (c) MgO (d) K_2S (e) Na_2SO_4 (f) $AgNO_3$ (g) K_2CO_3 (h) $Cu(NO_3)_2$ (i) $Fe(NO_3)_2$ (j) Fe_2O_3 $10x$ **(1)**

4. (a) $K_2CO_3 + CaBr_2 \rightarrow 2KBr + CaCO_3$

   (b) $Cu(NO_3)_2 + 2NaOH \rightarrow Cu(OH)_2 + 2NaNO_3$

   (c) $Al_2(SO_4)_3 + 6NaOH \rightarrow 3Al(OH)_3 + 3Na_2SO_4$

   (d) $2AgNO_3 + CaBr_2 \rightarrow 2AgBr + Ca(NO_3)_2$

   (e) $3Pb(NO_3)_2 + Al_2(SO_4)_3 \rightarrow 3PbSO_4 + 2Al(NO_3)_3$

For all: correct LHS (1) correct RHS (1) balance **(1)**

5. (a) $Al(27) + N(3 \times 14) + O(9 \times 16) = 213$ g = 1 mole **(1)**
   0.25 mole = 213 x 0.25 = 53.25 g **(1)**
   (b) $Cu(64) + S(32) + O(4 \times 16) + 5H_2O (5 \times 18) = 250$ g = 1 mole **(1)**
   1.50 moles = 1.50 x 250 = 375 g **(1)**

6. (a) 0.20 moles (That is the meaning of '0.20 M') **(1)**
   (b) If there are 0.20 moles in 1 dm^3 (1000 cm^3) there will be $0.20 \div 1000$ moles in 1.0 cm^3
   So there will be $25(0.20 \div 1000) = 0.005$ moles in 25 cm^3 **(2)**
   (c) From eqn each mole of NaOH reacts with HALF a mole of H_2SO_4
   So moles H_2SO_4 in 20 cm^3 = 0.5 x 0.005 = 0.0025 moles **(2)**
   (d) Moles H_2SO_4 = 0.0025 = volume x molarity x 10^{-3} = 20 x molarity x 10^{-3}
   Giving molarity = 0.125M **(2)**
   (e) M_r H_2SO_4 = 98. Mass = moles x M_r = 0.125 x 98 = 12.25 g/dm^3 ($g\ dm^{-3}$) **(2)**

F Ionic compounds

1. **(10)**

Element	Electronic structure of atom	Electronic structure of ion
(a) lithium	2.1	2
(b) oxygen	2.6	2.8
(c) aluminium	2.8.3	2.8
(d) potassium	2.8.8.1	2.8.8
(e) chlorine	2.8.7	2.8.8

2. Aluminium ion: 2e on inner ring **(1)**, 8e on next ring **(1)**

   chloride ion: 2e on inner ring **(1)**, 8e on next two rings **(1)**

3. **(12)**

Symbol for element	Electronic configuration of atom	Electronic configuration of ion	Symbol for ion
A	2.1	2	A^+
B	2.6	2.8	B^{2-}
C	2.7	2.8	C^-
D	2.8.2	2.8	D^{2+}
E	2.8.3	2.8	E^{3+}
F	2.8.8.1	2.8.8	F^+

4. (a) Na_2O < MgO < Al_2O_3 **(2)**

   (b) Al_2O_3 has $3+$ cation and $2-$ anions, so largest electrostatic attraction between the ions **(1)**

   MgO has $2+$ cation and $2-$ anions, so has a larger electrostatic attraction between the ions **(1)**

   than those in Na_2O, which contains only a $1+$ cation **(1)**

5. NaBr Na atom 2.8.1 shown transferring one electron to Br atom 2.8.18.7 to form one Na^+ and one Br^- ion

   $MgCl_2$ Mg atom 2.8.2 transferring one electron to each of two Cl atoms 2.8.7 to form one Mg^{2+} and two Cl^- ions

   AlF_3 Al atom 2.8.3 transferring one electron to each of three F atoms 2.7 to form one Al^{3+} and three F^- ions

   MgO Mg atom 2.8.2 transferring two electrons to an O atom 2.6 to form one Mg^{2+} and one O^{2-} ion

   Li_3N 3 Li atoms 2.1 transferring one electron to one N atom 2.5 to form three Li^+ ions and one N^{3-} ion

   K_2O 2 K atoms 2.8.8.1 transferring one electron to an O atom 2.6 to form two K^+ and one O^{2-} ion

   metal atom electronic structure correct **(1)**   non-metal atom electronic structure correct **(1)**

   electron transfer correct **(1)**

G Covalent substances

1. (a) (i) a structure in which very large numbers of atoms are all joined by covalent bonds **(1)**

    (ii) The substances are graphite and diamond:
    graphite has high electrical conductivity **(1)** because each carbon atom uses only 3 electrons in bonding **(1)** the remaining electrons are mobile and carry the current **(1)**
    diamond is hard **(1)** because each carbon atom forms 4 strong covalent bonds **(1)** so moving one atom with respect to the others requires considerable energy / force **(1)**

2. Sodium occurs just after neon in the Periodic Table **(1)** by losing one electron it achieves a stable inert gas structure **(1)** so forms a +1 ion **(1)**

    Chlorine occurs just before argon in the Periodic Table **(1)** by gaining one electron it achieves a stable inert gas structure **(1)** so forms a −1 ion **(1)**

    Carbon would need to lose or gain four electrons to achieve an inert gas structure (He or Ne) **(1)** both of these options (usually) require too much energy to be recouped by the formation of an ionic compound **(1)** so the chemistry of carbon is predominantly covalent **(1)**

    (Note: Ionic compounds of carbon containing the C^{4-} ion are known, for example in aluminium carbide, Al_4C_3, where the energy needed to form the C^{4-} ion is offset by the very large attractive forces between the Al^{3+} and C^{4-} ions.)

3. (a) 5x **(2)**

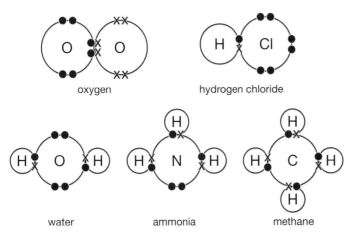

oxygen   hydrogen chloride

water   ammonia   methane

    (b) The central atoms share the number of electrons needed to give the compound a stable inert gas structure **(1)** sharing fewer electrons implies forming fewer bonds **(1)** and since forming bonds is exothermic the resulting molecules (such as OH, and CH_3) would be less stable **(1)** more electrons than the inert gas structure requires cannot be accommodated because no shells are available so molecules like NH_4 are not found **(1)**

H Metallic crystals

1. (a) Sketch similar to page 50 showing atoms in contact **(1)** regular arrangement **(1)**
   (b) The electrons **(1)** attract the nuclei **(1)** as a result of their opposite charges **(1)**

2. (a) Can be rolled / hammered **(1)** into sheets **(1)**
   (b) Layers **(1)** of metal atoms **(1)** can slide over one another **(1)**
   (c) Lithium is a smaller atom **(1)**

   so the forces of attraction between the nuclei and the electrons are greater **(1)**

   so it is harder to move one layer of metal atoms with respect to another **(1)**

3. (a) There is a mobile flux / sea **(1)** of electrons in a metal **(1)**

   and these carry a current when a voltage is applied **(1)**
   (b) Copper has more electrons **(1)** than sodium **(1)**
      so it carries a current with less resistance / more effectively **(1)**

I Electrolysis

1. (a) (i) $4OH^-(aq) - 4e^- \rightarrow 2H_2O(l) + O_2(g)$ (correct LHS **(1)** correct RHS **(1)**)

   (ii) $2H^+(aq) + 2e^- \rightarrow H_2(g)$ (eqn **(1)**, state symbols **(1)**)

   (b) $2Cl^-(aq) - 2e^- \rightarrow Cl_2(g)$ (eqn **(1)**, state symbols **(1)**)

   (c) $Cu^{2+}(aq) + 2e^- \rightarrow Cu(s)$ (eqn **(1)**, state symbols **(1)**)

   (d) (i) $2O^{2-} - 4e^- \rightarrow O_2(g)$ (eqn **(1)**, state symbols **(1)**)

   (ii) $Al^{3+} + 3e^- \rightarrow Al(l)$ (eqn **(1)**, state symbols **(1)**)

2. (a) So that the ions it contains **(1)** can move **(1)**

   (b) Negative **(1)** positive **(1)**

   (c) Lead is formed when lead ions **(1)** are reduced **(1)** when they gain two electrons **(1)**

   Bromine is formed when bromide ions **(1)** are oxidised **(1)** when they each lose one electron **(1)**

3. (a) In the copper wires and the graphite electrodes **(1)** mobile electrons carry the current **(1)**

   in the molten sodium chloride the chloride ions move to the anode **(1)** where they give up electrons **(1)**

   the sodium ions move to the cathode **(1)** where they gain electrons **(1)**

   (b) Cathode: $Na^+ + e^- \rightarrow Na$ **(2)**

   anode: $2Cl^- - 2e^- \rightarrow Cl_2$ **(2)**

   (c) (i) Oxidation at the anode **(1)** chloride ions lose electrons **(1)**

   (ii) Reduction at the cathode **(1)** sodium ions gain electrons **(1)**

4. (a) (i) cathode: $2H^+ + 2e^- \rightarrow H_2$ **(2)**   anode: $4OH^- - 4e^- \rightarrow 2H_2O + O_2$ **(2)**

   (ii) cathode: $Cu^{2+} + 2e^- \rightarrow Cu$ **(2)**   anode: $2Cl^- - 2e^- \rightarrow Cl_2$ **(2)**

   (iii) cathode: $Na^+ + e^- \rightarrow Na$ **(2)**   anode: $2Cl^- - 2e^- \rightarrow Cl_2$ **(2)**

   (iv) cathode: $Pb^{2+} + 2e^- \rightarrow Pb$ **(2)**   anode: $2Br^- - 2e^- \rightarrow Br_2$ **(2)**

   (b) Hydrogen: a test tube of the gas burns with a squeaky pop **(1)**

   Oxygen: relights a glowing splint **(1)**

   Sodium: silvery metal with a low melting point, globules of it float on the electrolyte **(1)**

   Chlorine: turns moist blue litmus paper red, then bleaches it **(1)**

   Lead: soft metal, marks paper **(1)**

   Bromine: reddish vapour / turns aqueous potassium iodide deep red-brown **(1)**

5. 1 tonne of Al $= 10^6 \div 27$ moles $= 3.7 \times 10^4$ moles **(1)**

   Each mole of Al requires 3 faradays of charge (because the Al ion has a +3 charge) **(1)**

   Number of faradays to produce 3.7×10^4 moles of Al $= 3 \times 3.7 \times 10^4 = 1.11 \times 10^5$ faradays **(2)**

6. $480\ dm^3$ of oxygen $= 480 \div 24$ moles $= 20$ moles of oxygen molecules **(1)**

   Each mole of oxygen molecules requires 4 faradays of charge: $2O^{2-} - 4e^- \rightarrow O_2(g)$ **(1)**

   Number of faradays of charge to form $480\ dm^3$ (20 moles) of O_2 molecules $= 4 \times 20 = 80$ faradays **(1)**

   Each mole of Al requires 3 faradays of charge (formed from a +3 ion) **(1)**

   Moles of Al formed $= 80 \div 3 = 26.7$ moles **(1)**

   Mass of Al formed = moles of Al formed x A_r of Al $= 26.7 \times 27 = 720.9$ g **(1)**

2 Chemistry of the elements

A The Periodic Table

1. (a) (i) C **(1)** Group 0 elements are monatomic / contain atoms only **(1)**

    (ii) A **(1)** Group 7 elements consist of molecules containing two identical atoms **(1)**

    (iii) B **(1)** because it is a molecule containing two different atoms **(1)**

    (iv) D **(1)** the product is water, which contains one O atom bonded to two H atoms **(1)**

  (b) (i) potassium atom 19 electrons, caesium ion 54 electrons **(2)**

    (ii) potassium atom 19 protons, caesium ion 55 protons **(2)**

    (iii) 1 electron **(1)**

2. (a) N and B **(2)**

  (b) G **(1)**

  (c) L **(1)**

  (d) L **(1)**

  (e) F **(1)**

  (f) K **(1)**

  (g) I **(1)**

  (h) J **(1)**

  (i) D **(1)**

  (j) M **(1)**

  (k) C and L, or I and G **(2)**

  (l) Period 1 **(1)**

  (m) Two of M, D, E, or J **(2)**

  (n) G with C, H or I **(2)**

  (o) Two of F, D, E, J or N **(2)**

B Group 1 elements

1. (a) Melting point decreases **(1)**

  (b) density increases **(1)**

2. (a) Metal floats / melts / moves on surface, bubbles of gas seen, purple flame **(any 3)**

  (b) (i) Hydrogen **(1)** (ii) sodium hydroxide **(1)**

  (c) Add litmus **(1)** turns blue **(1)** or phenolphthalein **(1)** turns pink **(1) or** add Universal Indicator **(1)** turns violet **(1)**

3. (a) (i) Lithium < sodium < potassium **(2)**

    (ii) Reaction involves electron loss **(1)** this is easier the larger the atom Li < Na < K **(1)** and easier when there is greater shielding is the case Li < Na < K **(1)**

  (b) Lithium does not melt / moves less rapidly on the surface / gives off gas less rapidly **(any 1)**

4. Silvery metal **(1)**, low melting point **(1)**, conducts electricity well **(1)**, soft/very malleable **(1)**, reacts violently with water **(1)**, to form hydrogen **(1)**, and an alkaline solution of its hydroxide **(1) (any 3)**

C Group 7 elements

1. sodium **(1)** Na⁺ **(1)** chlorine **(1)** Cl⁻ **(1)** neon **(1)** argon **(1)**

2. (a) Poor conductors of heat **(1)** and electricity **(1)** form negative ions **(1)** form acidic oxides **(1) (any 2)**

   (b) 7 electrons associated with each Cl **(1)** shared pair **(1)** one from each Cl **(1)**

3. (a) Hydrogen chloride **(1)**

   (b) (i) Hydrochloric acid **(1)**

   (ii) $HCl(g) + H_2O(l) \rightarrow H_3O^+(aq) + Cl^-(aq)$ **(3)**

   (c) To the solution in methylbenzene: dip litmus paper in it **(1)** no colour change **(1)**; add magnesium **(1)** no evolution of hydrogen **(1)**; check pH with a pH meter **(1)** pH = 7 **(1)**; add solid calcium carbonate **(1)** no carbon dioxide formed **(1)**
   dip electrodes connected in series to a battery and bulb **(1)** bulb fails to light / solution does not conduct electricity **(1)**

   (3 tests, 2 marks each)

4. (a) (i) Atomic number increases **(1)** so there are more electron shells **(1)**

   (ii) Elements react by gaining electrons **(1)** this becomes less favourable energetically as the atoms increases in size / have more electron shells shielding the outer shell **(1)**

   (b) Add an aqueous solution of an iodide **(1)** bubble in chlorine / add chlorine water **(1)** red-brown colour of (displaced) iodine seen **(1)**

   (c) $2I^-(aq) + Cl_2(aq)/(g) \rightarrow I_2(s)/(aq) + 2Cl^-(aq)$ species **(1)**, balance **(1)** state symbols **(1)**

5. (a) green **(1)** gas **(1)**

   (b) deep-red **(1)** liquid **(1)**

   (c) grey / metallic **(1)** solid **(1)**

6. Decreases **(1)** increases **(1)**

7. (a) (i) Redox **(1)** bromide ions have lost electrons **(1)** so have been oxidised **(1)** chlorine has gained electrons **(1)** so has been reduced **(1)**

   (ii) Colourless solution **(1)** becomes orange **(1)**

   (b) Bromine is a larger atom **(1)** with more shell of electrons shielding the outer shell **(1)** so it gains an electron less easily than chlorine **(1)**

8. (a) (i) A solvent having polar bonds / bonds between different atoms **(1)** and a shape which has no centre of symmetry / is not symmetrical **(1)** and has as a result a dipole / one end with a positive charge, the other with a negative charge **(1)**

   (ii) water **(1)**

   (b) Breaking of the H–Cl bond requires an input of energy **(1)** polar solvent is attracted to the resulting ions **(1)** and this releases energy greater than that needed to break the H–Cl bond / is more exothermic than the bond breaking **(1)** in a non-polar solvent no such exothermic reaction occurs **(1)**

   (c) (i) Add aqueous silver nitrate **(1)** acidified with nitric acid **(1)** white precipitate confirms chloride **(1)**

   (ii) Test with pH meter **(1)** a pH less than 7 **(1)** confirms that hydrogen ions are present **(1) or** add magnesium metal **(1)** to show that gas formed burns with a squeaky pop **(1)** confirms hydrogen ions present **(1)**

D Oxygen and oxides

1. (a) $2Mg + O_2 \rightarrow 2MgO$ **(2)**

   (b) $2C + O_2 \rightarrow 2CO$ **(2)**

2. (a) Diagram of two syringes **(1)** connected to tube packed with copper turnings **(1)** note initial volume of air **(1)** pass air back and forth over heated copper **(1)** until no further reduction in volume is apparent **(1)** note final volume of gas **(1)**

   (b) Volume of oxygen present = (85 – 68) = 17 cm³ **(1)** % oxygen = 100 x (17 ÷ 85) = 20% **(1)**

3. (a) An aqueous solution of gaseous acidic oxides / SO_2, NO_x **(1)** the oxides are formed during fossil fuel combustion **(1)** and are soluble in rainwater **(1)**

   (b) Acidification of waterways **(1)**, damage to living organisms **(1)**, enhanced corrosion of metals **(1)**, erosion of stonework **(1)** **(any 2)**

4. (a) (i) manganese(IV) oxide / dioxide **(1)**

   (ii) $2H_2O_2 \rightarrow 2H_2O + O_2$ (correct LHS **(1)** correct RHS **(1)** balance **(1)**)

5. (a) (i) Sulfur burns more vigorously **(1)** with a (pale) blue flame **(1)**

   (ii) $S + O_2 \rightarrow SO_2$ **(2)** sulfur dioxide **(1)**

   (b) (i) Sulfurous acid **(1)**
   (ii) Add indicator **(1)** show colour observed corresponds to a pH below 7 **(1)**
   (iii) Its oxide is acidic **(1)** characteristic of non-metals **(1)**

E Hydrogen and water

1. (a) (i) $Zn(s) + 2HCl(aq) \rightarrow ZnCl_2(aq) + H_2(g)$ (ii) $2Al(s) + 6HCl(aq) \rightarrow 2AlCl_3(aq) + 3H_2(g)$ (correct LHS **(1)** correct RHS **(1)** balance **(1)**)

   (b) (i) $H_2SO_4(aq) + Mg(s) \rightarrow MgSO_4(aq) + H_2(g)$ (ii) $H_2SO_4(aq) + Fe(s) \rightarrow FeSO_4(aq) + H_2(g)$ (correct LHS **(1)** correct RHS **(1)** balance **(1)**)

2. Turns anhydrous copper(II) sulfate **(1)** from white to blue **(1)** boils at 100 °C / freezes at 0 °C **(1)**
   or turns cobalt chloride papers **(1)** from blue to pink **(1)** boils at 100 °C / freezes at 0 °C **(1)**

3. (a) Water is a liquid, carbon dioxide is a gas at room temperature **(1)**

   water has a higher boiling point than carbon dioxide / water is easier to condense **(1)**

   so water can be prevented from escaping into the atmosphere more easily than carbon dioxide **(1)**

   (b) Advantages: need not be obtained from fossil fuels **(1)**; is renewable **(1)**; exhaust gases need not contribute to global warming **(1)**; has lower density than liquid fuels **(1)** **(any 2)**
   disadvantages: more difficult to handle / store; more expensive at present **(2)**

F Reactivity series

1. (a) Conduct electricity, malleable, ductile, sonorous, solid at room temperature **(any 3)**

    (b) Pink/brown coating **(1)**, colour of copper sulfate solution fades **(1)**, metal becomes smaller / dissolves **(1)**

    (c) (i) Hydrogen **(1)**

    (ii) millennium + sulfuric acid → millennium sulfate + hydrogen **(2)**

    (d) (i) Millennium (metal) **(1)**

    (ii) magnesium + millennium sulfate → magnesium sulfate + millennium **(3)**

    (e) (i) copper < millennium < magnesium **(2)**

    (ii) Magnesium is more reactive than millennium because it displaces it from solution **(1)**
    millennium is more reactive than copper because it displaces it from solution **(1)**
    copper is the least reactive of the three **(1)**

    (f) millennium oxide **(1)**

2. (a) (i) A substance which removes oxygen from a compound **(1)** by forming a (more) stable compound with it than that in the compound which is reduced **(1)**

    (ii) Yellow solid reacting / disappearing **(1)** shiny molten bead of lead formed **(1)**

    (iii) $PbO + C \rightarrow CO + Pb$ **(2)**

3. (a) Galvanising **(1)** because article still protected from rust even if the zinc coating is breached **(1)**

    (b) Oiling / greasing **(1)** because they will adhere to the moving parts of the chain **(1)**

    (c) Oiling / greasing **(1)** because other methods are not appropriate to the use made of the spade and the oil or grease can be easily re-applied when removed in use **(1)**

    (d) Painting **(1)** painting is decorative as well as protective / other methods incompatible with use on railings **(1)**

    (e) Sacrificial protection with zinc / magnesium block (not 'galvanising') **(1)** because structures too large to galvanise / operate in hostile environments in which paint would fail to protect **(1)**

4. (a) (i) Iron **(1)** it took the longest time to form 10 cm³ of hydrogen **(1)**

    (ii) Surface area of the samples of metal **(1)** the larger the surface area, the faster the rate **(1)**

    (b) Aluminium has an oxide coating on its surface **(1)** this prevents the acid reaching the metal **(1)**
    so the rate of reaction is slower than expected **(1)**

    (c) Oxygen and water are needed for rusting to occur **(1)** painting prevents both from reaching the surface of the iron **(1)**

    (d) Zinc took a shorter time than iron to produce 10 cm³ of gas **(1)** so it is more reactive than iron **(1)** and so corrodes in preference to it (thus protecting the iron from rusting) **(1)** even if the zinc coating is scratched **(1)** if the paint coating is chipped rusting can occur **(1)**

5. (a) The more reactive the metal **(1)** the greater the voltage produced **(1)**

    (b) Magnesium produces a larger voltage than iron **(1)** so is more reactive than iron **(1)** and therefore corrodes in preference to the iron, protecting it **(1)**

    (c) (i) zinc > iron > nickel > tin **(1)**

    (ii) Zinc produces the largest voltage **(1)** so it is the most reactive **(1)** the others have progressively smaller voltages in the order shown **(1)**

    (d) Iron produces a larger voltage than tin **(1)** so is more reactive **(1)**
    so if the tin coating is scratched it increases the tendency of the iron can to corrode **(1)**

    (e) (i) Nickel is the more reactive metal **(1)** so it can displace copper from solution **(1)**
    a pink / brown deposit of copper is seen **(1)** and the solution of copper sulfate becomes paler **(1)**

    (ii) $Ni^{2+}(aq) + Cu(s)$ **(2)**

G Tests for ions and gases

1.

Ion tested for	Test reagent	Observation
calcium / Ca^{2+}	**aqueous sodium hydroxide (1)**	white **(1) precipitate (1)**
copper(II) / Cu^{2+} iron(II) / Fe^{2+} iron(III) / Fe^{3+}	**aqueous sodium hydroxide**	**blue (1)** precipitate **green (1)** precipitate **red-brown (1)** precipitate
sulfate / SO_4^{2-}	**aqueous barium chloride (1)** **acidified with dilute hydrochloric** **acid (1)**	white **(1)** precipitate
chloride ions / Cl^- bromide ions / Br^- iodide ions / I^-	**aqueous silver nitrate (1)** **acidified with dilute nitric acid (1)**	**white (1)** precipitate **pale yellow / cream (1)** precipitate **yellow (1)** precipitate

2.   (a) yes: silver nitrate **(1)**   (b) no precipitate **(1)**   (c) no precipitate **(1)**   (d) yes: copper(II) hydroxide **(1)**

3 Organic chemistry

A Alkanes

1. (a) A compound with the same molecular formula as another **(1)** but with a different structure **(1)**

   (b)

   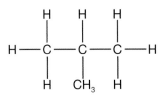

   butane

   each correct formula **(2 x 1)** correct name for butane **(1)**

2. (a) $C_3H_8 + 5O_2 \rightarrow 3CO_2 + 4H_2O$ **(2)**

   (b) $C_3H_8 + 3\frac{1}{2}O_2 \rightarrow 3CO + 4H_2O$ **(2)**

3. (a) C_nH_{2n+2} **(1)**

   (b) C_5H_{12} **(1)**

4. (a) In the presence of UV light / sunlight **(1)**

   (b) $CH_4 + Br_2 \rightarrow CH_3Br + HBr$ **(2)**

B Alkenes

1. (a) (i) C_nH_{2n} **(1)**

   (ii) names **(2)** formulae **(2)**

   but-1-ene   but-2-ene

   (iii) name **(1)** formula **(1)**

   2-methylpropene

2. (a) Shake **(1)** with bromine water **(1)**

   (b) Hexene – orange solution **(1)** turns colourless **(1)**

   Hexane – organic layer becomes orange **(1)**

C Ethanol

1. (a) (i) sugar + yeast → ethanol + carbon dioxide (+ heat energy) **(4)**

    (ii) Yeast is killed by high concentration of ethanol **(1)**

    (iii) Fractional **(1)** distillation **(1)**

    (iv) Boiling point **(1)**

    (b) (i) steam / water **(1)**

    conditions: 70 atmospheres pressure **(1)**, 300 °C, **(1)** phosphoric acid catalyst **(1)**

    (ii) Pure ethanol formed or no need for fractional distillation **(1)**; continuous process **(1)**; cheaper **(1)** **(any 2)**

    (iii) As a fuel / petrol substitute **(1)** saves money on imported fuels **(1)**

    (c) (i) Hot **(1)** clean **(1)** flame

    (ii) $C_2H_5OH + 3O_2 \rightarrow 2CO_2 + 3H_2O$ **(2)**

    (iii) Solvent / ester manufacture / chemical feedstock **(any 1)**

2. (a) (i) Powdered aluminium oxide **(1)** (ii) ethene **(1)** (iii) ceramic wool soaked in ethanol **(1)**

    (b) Dehydration **(1)**

    (c) Add bromine water **(1)** orange solution **(1)** becomes colourless **(1)**

D Carboxylic acids

1. (a) methanoic acid, (b) ethanoic acid, (c) propanoic acid **(3)**

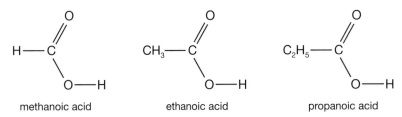

methanoic acid   ethanoic acid   propanoic acid

2. (a) An acid which is partially ionised **(1)** in aqueous solution **(1)**

   (b) $CH_3COOH(aq) \rightleftharpoons CH_3COO^-(aq) + H^+(aq)$

   $\rightleftharpoons$ (symbol **(1)** species on RHS correct **(2)**)

3.

   correct groups **(1)** correct molecular shape **(1)**

4. (a) (i) Ethanoic acid **(1)** ethanol **(1)**
       (ii) concentrated **(1)** sulfuric acid **(1)**

   (b) source of heat **(1)** suitable flask **(1)** condenser **(1)** water in/out correct **(1)**

4 Physical chemistry

A Acids, alkalis and salts

1. (a) (i) To measure a variable volume of solution **(1)**
   (ii) Vessel in which titration is carried out **(1)**
   (iii) To measure a fixed volume of solution **(1)**
   (b)   To show when pH = 7 / neutralisation complete **(1)**

2. (a) (i) $HCl(aq) \rightarrow H^+(aq) + Cl^-(aq)$ **(1)**
   (ii) $CH_3COOH(aq) \rightleftharpoons CH_3COO^-(aq) + H^+(aq)$ equation **(1)** equilibrium arrows **(1)**
   (b) Weak acid is partially / incompletely ionised in solution **(1)** strong acid is fully ionised in solution **(1)**
   (c) Hydrochloric **(1)**

3. acidic < 7 **(1)**, alkaline >7 **(1)**

4. (a) (i) Litmus: acidic red, alkaline blue **(1)**
   (ii) Phenolphthalein: acidic colourless, alkaline pink **(1)**
   (iii) Methyl orange: acidic red, alkaline yellow **(1)**
   (b) (i) red **(1)** (ii) blue **(1)**

5. $H^+(aq) + OH^-(aq) \rightarrow H_2O(l)$ (H^+ = **(1)**, OH^- = **(1)**, (aq) = **(1)**)

6. **(6)**

Acid	Alkali	Salt formed
nitric	sodium hydroxide	sodium nitrate
sulfuric	sodium hydroxide	sodium sulfate
hydrochloric	sodium hydroxide	sodium chloride
nitric	potassium hydroxide	potassium nitrate
sulfuric	potassium hydroxide	potassium sulfate
hydrochloric	potassium hydroxide	potassium chloride

7. (i) $Mg + H_2SO_4 \rightarrow MgSO_4 + H_2$ **(3)**
   (ii) $CuO + 2HNO_3 \rightarrow Cu(NO_3)_2 + H_2O$ **(3)**
   (iii) $ZnCO_3 + 2HCl \rightarrow ZnCl_2 + H_2O + CO_2$ **(4)**

8. Soluble: sodium bromide, ammonium sulfate, potassium carbonate and lead nitrate
   Insoluble: silverchloride barium carbonate
   All correct = **(6)**, one wrong = **(4)**, two wrong = **(2)**, three wrong = **(1)**

9. (a) $CuO(s) + H_2SO_4(aq) \rightarrow CuSO_4(aq) + H_2O(l)$ species **(2)** state symbols **(1)**
   (b) $2H^+(aq) + O^{2-}(s) \rightarrow H_2O(l)$ ions **(2)** state symbols **(1)**

10. (a) B **(1)** $CuO + H_2SO_4 \rightarrow CuSO_4 + H_2O$ **(3)**

(b) A **(1)** $KOH + HCl \rightarrow KCl + H_2O$ **(3)**

(c) C **(1)** $AgNO_3 + NaBr \rightarrow AgBr + NaNO_3$ **(3)**

(d) A **(1)** $HBr + NaOH \rightarrow NaBr + H_2O$ **(3)**

(e) B **(1)** $ZnO + 2HCl \rightarrow ZnCl_2 + H_2O$ **(3)**

(f) C **(1)** $BaCl_2 + K_2CO_3 \rightarrow BaCO_3 + 2KCl$ **(3)**

(g) C **(1)** $Pb(NO_3)_2 + 2NaI \rightarrow PbI_2 + 2NaNO_3$ **(3)**

(h) B **(1)** $CaCO_3 + 2HNO_3 \rightarrow Ca(NO_3)_2 + H_2O + CO_2$ **(3)**

(i) B **(1)** $Zn + 2HCl \rightarrow ZnCl_2 + H_2$ **(3)**

11. Add copper(II) oxide **(1)** to dilute sulfuric acid **(1)**
warm **(1)** make sure excess copper(II) oxide present **(1)**
filter **(1)** evaporate filtrate to about one-third of its volume **(1)**
leave to cool **(1)** decant / filter solution from crystals **(1)**

12. **(8)**

Solutions mixed	Precipitate formed? (Yes / No)	Name of precipitate
sodium chloride and silver nitrate	yes	silver chloride
sulfuric acid and barium nitrate	yes	barium sulfate
sodium carbonate and calcium nitrate	yes	calcium carbonate
potassium hydroxide and calcium chloride	no	N/A

13. Nitrates **(1)** sodium / potassium salts **(1)**

B Energetics

1. Weaker **(1)** stronger **(1)**

2. (a) x-axis – *time* or *progress of reaction* **(1)**   y-axis – *energy* **(1)**
   correct endothermic profile, with products at higher energy than reagents **(1)**
   activation energy: single-headed arrow pointing upwards from reactants to peak of curve **(1)**
   heat change: single-headed arrow pointing upwards from reactants to products **(1)**
   (Deduct 1 mark if double-headed arrows are used.)
   (b) Photosynthesis / thermal decomposition of calcium carbonate **(any 1)**

3. (a) (i) –Q **(1)** (ii) the energy change has the same magnitude **(1)** but its sign changes **(1)**
   (b) (i) Blue solid becomes white **(1)** steam / water vapour formed **(1)**
   (c) Endothermic L → R **(1)** because the reaction in this direction requires heat in order to take place **(1)**

4. (a) Energy **(1)**
   (b) (i) Energy / enthalpy change **(1)** (ii) activation energy **(1)**

5. (a) 2640 – 3338 = - 698 kJ/mol **(2)**

   (b) Exothermic **(1)** more energy given out when bonds formed than when bonds broken **(1)**

6. (a) Bonds broken: 2C–C   696   bonds formed: 8C=O   –5944
                       12C–H   4944   12O–H   –5556
                       7O=O   3472   –11500
                       total   9112 Δ**H** = –2388 kJ mol^{-1}

   (b) Bonds broken: 1C–C   348   bonds formed: 4C=O   –2972
                       5C–H   2060   6O–H   –2778
                       1C–O   360
                       1O–H   463   –5750
                       3O=O   1448
                       total   4719 Δ**H** = –1031 kJ mol^{-1}

   (c) Bonds broken: 4C–H   1648   bonds formed: 2C=O   –1486
                       4O–H   1852   4H–H   –1744
                                                                     –3230
                       total   3500 Δ**H** = +270 kJ mol^{-1}

   (d) Bonds broken: 1C=C   612   bonds formed: 1C–C   –348
                       1H–H   436   2C–H   –824
                                                                     –1172
                       total   1048 Δ**H** = –124 kJ mol^{-1}

   (e) Bonds broken: 2C–C   696   bonds formed: C=C   –612
                                                            Δ**H** = +84 kJ mol^{-1}

7. (a) $CH_3OH + 1\frac{1}{2}O_2 \rightarrow CO_2 + 2H_2O$ (correct LHS **(1)** correct RHS **(1)** balance **(1)**)

(b) moles of methanol = $0.21 \div 32$

    = 6.56×10^{-3} mol **(2)**

enthalpy change = $100 \times 4.2 \times 11.1$

= -4662 J / -4.66 kJ **(2)**

molar enthalpy change = $4.66 \div 6.56 \times 10^{-3}$

= -710 kJ/mol **(1)**

8. (a) (i) mol Cu^{2+} = $200 \times 1.0 \times 10^{-3}$ = 0.20 mol **(1)**

    (ii) mol Fe = $7.0 \div 28$ = 0.25 mol **(1)**

    (iii) iron **(1)**

(b) enthalpy change = $200 \times 4.2 \times (41.5 - 18.5)$ = $-19\,320$ J for 0.20 mol Cu^{2+} reacted **(1)**

    molar enthalpy change = $19\,320 \div 0.20$ = $-96\,600$ J mol^{-1} / -96.6 kJ mol^{-1} answer **(1)** correct units **(1)**

(c) no heat losses **(1)**; specific heat capacity of copper sulfate solution is the same as that of water **(1)**; density of copper sulfate solution is the same as that of water **(1)**; heat capacities of iron used and copper formed neglected **(1)** **(any 2)**

9. (a) molar mass of ammonium chloride, NH_4Cl, = $14 + (4 \times 1) + 35.5$ = 53.5 g mol^{-1} **(1)**

mol ammonium chloride used = $5.40 \div 53.50$ = 0.1009 mol **(1)**

enthalpy change = $100 \times 4.2 \times (17.10 - 20.55)$ = -1449 J **(1)**

molar enthalpy change = $-1449 \div 0.1009$ = $14\,361$ J mol^{-1} / 14.36 kJ mol^{-1} **(1)**

(b) endothermic **(1)** because the temperature decreases during the reaction **(1)**

10. mol of both reactants = $50 \times 0.05 \times 10^{-3}$ = 2.5×10^{-3} mol **(1)**

enthalpy change = $100 \times 4.2 \times 0.31$ = 130.2 J **(1)**

molar enthalpy change = $130.2 \div 2.5 \times 10^{-3}$ = -52.08×10^{3} J mol^{-1} / -52.1 kJ mol^{-1} **(1)**

C Rates of reaction

1. (a) Rate increases **(1)** average energy increases at higher temperature **(1)** so more particles have energy in excess of the activation energy **(1)** and the collision rate also increases **(1)**

   (b) Rate decreases **(1)** collisions must occur for reaction to take place **(1)** a decrease in concentration implies fewer particles in a given volume **(1)** so there must be fewer collisions **(1)**

   (c) Rate increases **(1)** reaction can only occur on the surface **(1)** reducing the particle size increases surface area **(1)** so there are more effective collisions **(1)**

   (d) Rate increases **(1)** catalyst provides an alternative reaction pathway of lower activation energy **(1)** so more particles have energies greater than the (lower) activation energy **(1)** so there are more effective collisions **(1)**

2. (a) (i) Line 'A': choice of scales **(2)** axes labelled **(2)** points **(1)** line of best fit **(1)**

   (ii) I. Line 'B' steeper gradient than A **(1)** final mass 70.2 g **(1)**

   II. Line 'C' same gradient as A **(1)** final mass 70.6 g **(1)**

   (iii) Increased temperature leads to increased collision rate **(1)** and to an increase in average energy **(1)** so there are more effective collisions **(1)**

   (b) (i) Gradient of the line at time $t = 0$ minutes $= -0.32$ g min^{-1} **(3)**

   Gradient of the line at time $t = 5$ minutes $= -0.045$ g min^{-1} **(3)**

   (ii) Reaction faster at $t = 0$ minutes **(1)** because gradient greater at $t = 0$ than at $t = 5$ minutes **(1)**

   (iii) (rate at $t = 0$ minutes) / (rate at $t = 5$ minutes) $= -0.32 / -0.045$ (1) $= 7.1$ **(1)**

D Equilibria

1. (a) Right **(1)** because fewer moles on RHS **(1)**
   (b) Right **(1)** because equilibrium position shift to exothermic side **(1)**
   (c) Right **(1)** because forward reaction is endothermic **(1)**
   (d) Left **(1)** because fewer moles on LHS **(1)**
   (e) Left **(1)** because back reaction is exothermic **(1)**

5 Chemistry in society

A Extraction and uses of metals

1. (a) (i) $2O^{2-} - 4e^- \rightarrow O_2$ **(2)**

    (ii) $Al^{3+} + 3e^- \rightarrow Al$ **(2)**

    (b) (i) Cryolite **(1)**

    (ii) Its melting point is too high **(1)** to allow it to be melted economically **(1)**
    [common error is just to give the first point]

    (iii) So that the ions **(1)** [common error 'electrons'] are mobile / can move **(1)**

    (c)   Anode **(1)** burns away in the oxygen produced there **(1)**

2. Cheap / plentiful electricity **(1)**
    then **(1) each for any 2 of the following:** near a deep-water port / on an area of flat or cheap land / good transport links / available workforce

3.

Use	Related property
aircraft/window frames **(1)**	corrosion resistance **(1)**
cookware **(1)**	good thermal conductivity **(1)**
power cables [common error 'electrical wires'] **(1)**	good electrical conductivity **(1)**
foil **(1)**	malleability **(1)**

(any 3 uses and related properties)

4. (a) (i) 100 tonnes Al $= 10^8$ g $= 10^8 \div 27$ mol $= 3.70 \times 10^7$ mol Al **(1)**
    1 mol Al requires: ½ mol $Al_2O_3 = 3.70 \times 10^7 \div 2$ mol $Al_2O_3 = 1.85 \times 10^7$ mol **(1)**
    M_r $Al_2O_3 = 102$, so mass $Al_2O_3 = 102 \times 1.85 \times 10^7$ g $= 1.89 \times 10^9$ g / 1890 tonnes **(1)**

    (ii) For each mole of Al formed ¾ mol O_2 are formed **(1)**
    so mol $O_2 = ¾ \times 1.85 \times 10^7$ mol $= 1.39 \times 10^7$ mol **(1)**
    volume of O_2 formed at rtp $= 24 \times 1.39 \times 10^7$ dm³ $= 3.34 \times 10^8$ dm³ **(1)**

5. Aluminium is more reactive than carbon / its oxide cannot be reduced by carbon **(1)**
    iron is less reactive than aluminium and its oxide can be reduced by carbon **(1)**
    gold is very unreactive so is found 'native' **(1)**

6. (a) United Kingdom **(1)**
    (b) Lack of sufficient electricity **(1)**; poor transport links **(1)**; inadequate port facilities **(1) (any 2)**

7. (a) **(7)**

Raw material	Purpose
iron ore	source of the iron
coke	fuel and (source of) reducing agent
limestone	to form slag / remove impurities

(b) Hot **(1)** air **(1)**

8. (a) (i) $C(s) + O_2(g) \rightarrow CO_2(g)$ **(1)**

    (ii) $CO_2(g) + C(s) \rightarrow 2CO(g)$ **(2)**

    (iii) $Fe_2O_3(s) + 3CO(g) \rightarrow 2Fe(l) + 3CO_2(g)$ **(3)**

(b) (i) Calcium carbonate / limestone **(1)**

    (ii) So that it can be run off periodically **(1)** to make the manufacturing process continuous **(1)**

9. The ore and carbon are both solids so reaction occurs only on the surface / is inefficient **(1)**
but carbon monoxide is a gas and contact between it and the ore is better **(1)**
so the reaction involving carbon monoxide will be faster **(1)**

AM 10. (a) (i) Oxygen **(1)** is blown over the molten iron **(1)**

    (ii) Carbon is removed **(1)** makes the metal less brittle **(1)**

(b) (i) To produce steels with desirable properties e.g. stainless / stronger / springier / harder **(1)**

    (ii) one from: nickel, chromium, manganese, vanadium **(1)**

    (iii) The mixture is known as an alloy **(1)**

11. Haematite **(1)**

12. (a) (i) 4% **(1)**

    (ii) It is brittle **(1)**

(b) (i) Oxygen **(1)** blown through/ over the molten iron **(1)**

    (ii) About 1% **(1)**

B Crude oil

1. Crude oil is a mixture of compounds containing **carbon** and **hydrogen (1)** only. These compounds are called **hydrocarbons (1)**. The mixture can be separated by the process of **fractional (1) distillation (1)**. This process separates the mixture into **fractions (1)** containing several compounds having similar **boiling points (1)**.

2. (a) Boiling point increases **(1)** it flows less easily / more viscous **(1)**

3. (a) (i) carbon dioxide **(1)** + water
   (ii) carbon monoxide **(1)** + water   water correct in both **(1)**
   (b) To ensure no fuel is wasted **(1)** carbon monoxide formed is toxic **(1)**

4. (a) $N_2 + O_2 \rightarrow 2NO$ / $N_2 + 2O_2 \rightarrow 2NO_2$ **(2)**
   (b) The bond between the nitrogen atoms in N_2 is strong **(1)**
   so much energy is needed to break it / start the reaction **(1)**
   sufficiently high temperatures are reached in a car engine **(1)**

5. (a) (i) Alkanes (1) its formula fits their general formula C_nH_{2n+2} **(1)**
   (ii) No change in colour **(1)**
   because $C_{10}H_{22}$ is an alkane / is saturated / has no C=C **(1)**
   but organic layer becomes orange because (non-polar) Br_2 is more soluble in the (non-polar) alkane than in water **(1)**
   (b) (i) $C_{10}H_{22} \rightarrow C_4H_8 + C_6H_{14}$ both products **(2)**
   (ii) Shorter-chain alkanes of greater commercial value / alkenes can be used to make polymers **(1)**
   (c) (i) Heating needed to vaporise the alkanes in the paraffin **(1)**
   so the vapour passes over the catalyst **(1)**
   (ii) An immiscible liquid has distilled over **(1)**
   it could be one of the liquid alkanes present in the paraffin / a liquid alkene **(1)**

C Synthetic polymers

1. (a) (i) poly(ethene) $-CH_2-CH_2-CH_2-CH_2-CH_2-CH_2-$ **(1)**

   (ii) poly(chloroethene) $-CHCl-CH_2-CHCl-CH_2-CHCl-CH_2-$ **(2)**

   (b) (i) $CHCl=CCl_2$ **(1)**

   (ii) $CHCH_3=CHCH_3$ **(1)**

2. (a)   Plastic bags **(1)**, sheeting **(1)**, pipes **(1)** **(any 1)**

   (b)   Window frames **(1)**, guttering **(1)**, electrical insulation **(1)** **(any 1)**

   (c)   Crates **(1)**, rope **(1)**, pipes **(1)** **(any 1)**

3. Packaging **(1)** non-biodegradable **(1)** create toxic fumes / smoke when incinerated to dispose of them **(1)**

4. (a) Addition polymers can be made from only a single monomer **(1)**

   condensation polymers must be made from two different monomers **(1)**

   addition polymerisation occurs with no loss of material / the monomer and polymer have the same empirical formula **(1)**

   condensation polymerisation involves loss of a small molecule as the polymer forms **(1)**

   (b)   Nylon: ropes **(1)**, clothing **(1)**, engineering uses **(1)** **(any 1)**

   Terylene: fabric **(1)**, containers **(1)**, ropes **(1)** **(any 1)**

D The manufacture of some important chemicals

1. (a)   Finely divided iron **(1)**
    (b) (i) A dynamic equilibrium exists when the composition of a reaction mixture is unchanged with time **(1)** and when the forward and back reactions are proceeding at the same rate **(1)**
       (ii) (I) An increase in pressure increases the yield of ammonia **(1)** because there are fewer moles on the right-hand / product side of the equilibrium **(1)** and the equilibrium position shifts to this side to reduce the pressure **(1)**
           (II) A decrease in temperature increases the yield of ammonia **(1)** because the reaction is exothermic left to right **(1)** and the equilibrium position shifts to the right-hand / products side to increase the temperature **(1)**
       (iii) 450 °C **(1)** 250 atmospheres **(1)**
       (iv) Lower temperatures (less than 450 °C) would increase the yield **(1)** but the rate would be uneconomically slow **(1)**
           450 °C is a compromise temperature **(1)** giving an acceptable yield at a sufficiently rapid rate **(1)**

2. (a) Air **(1)** sulfur **(1)** water **(1)**
    (b) (i) they remain constant **(1)**
       (ii) they are the same **(1)**
    (c) (i) 450 °C **(1)**
       (ii) Because the reaction is exothermic a lower temperature would give a greater yield **(1)** / but the reaction is too slow at lower temperatures **(1)**
           so a compromise temperature of 450 °C is used to achieve a good yield at a reasonable rate **(1)**
       (iii) 2–5 atm / a little above atmospheric **(1)**
       (iv) The profit made on the extra yield would not be sufficient **(1)** to offset the additional costs of using a higher pressure **(1)**
       (v) vanadium(V) oxide / vanadium pentoxide **(1)**
           The catalyst provides an alternative reaction pathway of lower activation energy **(1)** so the same rate of reaction / establishment of equilibrium can be achieved at a lower temperature (than without a catalyst) **(1)** and the energy saving more than pays for the catalyst **(1)**

Note: There is an environmental advantage too – less fuel is used so reducing emissions of carbon dioxide.

    (d) Manufacture of fertiliser **(1)**, paint **(1)**, detergents **(1)**, any other valid **(1)** **(any 2)**
    (e) If added directly to water a very exothermic reaction occurs **(1)** the dense mist (of sulfuric acid) formed is slow to settle **(1)** instead, the acid is added to 98% sulfuric acid **(1)** forming more concentrated acid which is diluted with water to give the 98% (concentrated) acid **(1)**

3. (a) 66 ± 2%
    (b) 300 tonnes SO_2 = $(300 \times 10^6) \div 64$ mol = 4.69×10^6 mol **(1)**
        from the equation for a 100% yield: mol SO_3 = mol SO_2 = 4.69×10^6 mol **(1)**
        mass SO_3 = 4.69×10^6 mol $\times$ 80 g **(1)** = 3.75×10^8 g / 375 tonnes **(1)**
        but the yield is only 66%, so actual yield = 3.75×10^8 g / 375 tonnes $\times$ 0.66 = 2.48×10^8 g / 248 tonnes **(1)**

4. (a) (i)   brine **(1)**

(ii)   chlorine **(1)**

(iii) hydrogen **(1)**

(iv) titanium anode **(1)**

(v)   porous / semipermeable membrane **(1)**

(vi) nickel or steel cathode **(1)**

(vii) and (viii) brine **(1)** and sodium hydroxide solution **(1)**

(b) (i) $2H^+(aq) + 2e^- \rightarrow H_2$ **(2)**

(ii) $2Cl^- - 2e^- \rightarrow Cl_2$ **(2)**

(c) The H^+ and OH^- ions from the water are originally present in equal amounts $H_2O \rightarrow H^+ + OH^-$ **(1)**

during the electrolysis H^+ ions are removed as hydrogen gas at the cathode **(1)**

so OH^- ions are in excess in the region of the cathode, so the solution becomes alkaline **(1)**

(d) (i)   Manufacture of paper **(1)**, ceramics **(1)**, soaps **(1)**, detergents **(1)** **(any 2)**

(ii) Bleach **(1)**, killing bacteria in water supplies **(1)**, manufacture of PVC and hydrochloric acid **(1)** **(any 2)**

(e) From the equations, 2 mol of electrons are required to form 1 mol of chlorine **(1)**

2 mol of electrons remove 2 mol of H^+ ions so 2 mol of OH^- ions are also formed **(1)**

mol OH^- ions formed = 2 x mol Cl_2 formed = 40 mol **(1)**

Appendix

The Periodic Table

Group

Period	1	2												3	4	5	6	7	0
1	1 **H** Hydrogen 1																		4 **He** Helium 2
2	9 **Li** Lithium 3	9 **Be** Beryllium 4												11 **B** Boron 5	12 **C** Carbon 6	14 **N** Nitrogen 7	16 **O** Oxygen 8	19 **F** Fluorine 9	20 **Ne** Neon 10
3	23 **Na** Sodium 11	24 **Mg** Magnesium 12												27 **Al** Aluminium 13	28 **Si** Silicon 14	31 **P** Phosphorus 15	32 **S** Sulfur 16	35.5 **Cl** Chlorine 17	40 **Ar** Argon 18
4	39 **K** Potassium 19	24 **Ca** Calcium 20	45 **Sc** Scandium 21	48 **Ti** Titanium 22	51 **V** Vanadium 23	52 **Cr** Chromium 24	55 **Mn** Manganese 25	56 **Fe** Iron 26	59 **Co** Cobalt 27	59 **Ni** Nickel 28	63.5 **Cu** Copper 29	65 **Zn** Zinc 30	70 **Ga** Gallium 31	73 **Ge** Germanium 32	75 **As** Arsenic 33	79 **Se** Selenium 34	80 **Br** Bromine 35	84 **Kr** Krypton 36
5	86 **Rb** Rubidium 37	88 **Sr** Strontium 38	89 **Y** Yttrium 39	91 **Zr** Zirconium 40	93 **Nb** Niobium 41	96 **Mo** Molybdenum 42	99 **Tc** Technetium 43	101 **Ru** Ruthenium 44	103 **Rh** Rhodium 45	106 **Pd** Palladium 46	108 **Ag** Silver 47	112 **Cd** Cadmium 48	115 **In** Indium 49	119 **Sn** Tin 50	122 **Sb** Antimony 51	128 **Te** Tellurium 52	127 **I** Iodine 53	131 **Xe** Xenon 54
6	133 **Cs** Caesium 55	137 **Ba** Barium 56	139 **La** Lanthanum 57	179 **Hf** Hafnium 72	181 **Ta** Tantalum 73	184 **W** Tungsten 74	186 **Re** Rhenium 75	190 **Os** Osmium 76	192 **Ir** Iridium 77	195 **Pt** Platinum 78	197 **Au** Gold 79	201 **Hg** Mercury 80	204 **Tl** Thallium 81	207 **Pb** Lead 82	209 **Bi** Bismuth 83	210 **Po** Polonium 84	210 **At** Astatine 85	222 **Rn** Radon 86
7	223 **Fr** Francium 87	226 **Ra** Radium 88	227 **Ac** Actinium 89															

key

00 — relative atomic mass	metal
Xx — symbol	semi-metal
Name 00 — atomic number	non-metal
	inert gas